Gender Research in Natural Resource Management

The aim of this book is to highlight the role that gender research can play in understanding natural resource management and rural development issues in the Middle East and North Africa region and demonstrate how hands-on training and mentoring can be used as a capacity enhancement approach. It presents findings of four research teams, working in three countries: Algeria, Lebanon and Morocco.

The approach was to build the capacity of national teams through hands-on field research, mentoring and technical support, which is different from traditional training. The chapters present the results of the case studies used as learning platforms for the teams and reflections on this approach. The results demonstrate the capacity and skills that the teams acquired. These teams were multi-disciplinary and included social scientists as well as specialists in water management, livestock production and rangelands and agronomy. The book provides grounded, empirical examples for MENA academics, practitioners and development students concerned about ensuring gender-balanced rural development.

Malika Abdelali-Martini is a Socio-Economist and Rural Development and Gender Research Specialist in the Social, Economic and Policy Research Program at the International Center for Agricultural Research in the Dry Areas (ICARDA) based in Amman, Jordan.

Aden Aw-Hassan is Director of the Social, Economic and Policy Research Program at the International Center for Agricultural Research in the Dry Areas (ICARDA) based in Amman, Jordan.

Earthscan Studies in Natural Resource Management

Contested Forms of Governance in Marine Protected Areas
A study of co-management and adaptive co-management
Natalie Bown, Tim S. Gray and Selina M. Stead

Adaptive Collaborative Approaches in Natural Resource Management
Rethinking participation, learning and innovation
Edited by Hemant R. Ojha, Andy Hall and Rasheed Sulaiman V

Integrated Natural Resource Management in the Highlands of Eastern Africa
From concept to practice
Edited by Laura Anne German, Jeremias Mowo, Tilahun Amede and Kenneth Masuki

For more information on books in the Earthscan Studies in Natural Resource Management Series, please visit the series page on the Routledge website: http://www.routledge.com/books/series/ECNRM/

Gender Research in Natural Resource Management

Building capacities in the Middle East and North Africa

Edited by Malika Abdelali-Martini and Aden Aw-Hassan

Routledge
Taylor & Francis Group
LONDON AND NEW YORK

earthscan
from Routledge

First published 2014 by Routledge

2 Park Square, Milton Park, Abingdon, Oxfordshire OX14 4RN
52 Vanderbilt Avenue, New York, NY 10017

Routledge is an imprint of the Taylor & Francis Group, an informa business

First issued in paperback 2018

British Library Cataloguing-in-Publication Data
A catalogue record for this book is available from the British Library

Library of Congress Cataloging-in-Publication Data
Gender research in natural resource management : building capacities in the Middle East and North Africa / edited by Malika Abdelali-Martini and Aden Aw-Hassan.
pages cm. – (Earthscan studies in natural resource management)
Includes bibliographical references and index.
1. Women in development–Developing countries. 2. Rural women–Developing countries. 3. Women in agriculture–Developing countries. 4. Sustainable development–Developing countries. 5. Rural development–Environmental aspects–Middle East. 6. Rural development–Environmental aspects–North Africa. I. Abdelali-Martini, Malika. II. Aw-Hassan, Aden.
HQ1240.5.D44G4636 2014
305.409172'4–dc23
2013026013

ISBN: 978-0-415-72852-2 (hbk)
ISBN: 978-0-367-17335-7 (pbk)

Typeset in Bembo
by FiSH Books Ltd, Enfield

ICARDA International Center for Agricultural Research in the Dry Areas

IDRC | CRDI

Canada

The research presented in this publication was carried out with the financial assistance of Canada's International Development Research Centre.

Contents

Illustrations

Figures

Tables

Contributors

Malika Abdelali-Martini is a Socio-Economist and Rural Development and Gender Research Specialist in the Social, Economic and Policy Research Program at the International Center for Agricultural Research in the Dry Areas (ICARDA) based in Amman, Jordan.

Lina Abou-Habib is an Executive Director, The Collective for Research and Training on Development – Action (CRTD.A), Beirut, Lebanon and Programme Advisor for the Women Learning Partnership, Washington DC.

Habbouba Aoun is an Instructor at the Faculty of Health Sciences University of Balamand, Lebanon.

Aden Aw-Hassan is Director of the Social, Economic and Policy Research Program at the International Center for Agricultural Research in the Dry Areas (ICARDA) based in Amman, Jordan.

Mohamed Benidir is an Agricultural Engineer, Unité de recherche Sétif Institut National de la Recherche Agronomique (INRA), Algeria.

Abdelmadjid Benmebarek is a Professor/Researcher, École Nationale Supérieure d'Agronomie (ENSA), Algiers, Algeria.

Mohamed Larbi Cherfaoui is a Researcher, Institut National de la Recherche Agronomique (INRA), Algiers, Algeria.

Liz Fajber is a Senior Social Development Advisor, Department for International Development (DFID), London, UK.

Naima Faouzi is an Independent Consultant, Casablanca, Morocco.

Maria Fernandez is an Independent Consultant, Peru.

Souhila Fodil is Ingénieur en Agropastoralisme, Institut Technique Moyen d'Agronomie (ITMA), Djelfa, Algeria.

Asmae Harrad is a Student/Researcher, École Nationale d'Agriculture de Meknès (ENA), Meknes, Morocco.

Soumaya Ibrahim Huber is an Independent Consultant, Cairo, Egypt.

Johann Huguenin is Cadre Scientifique Recherche/Formation, Gestion de la Recherche, Centre de Coopération Internationale en Recherche Economique pour le Développement (CIRAD), Montpellier, France.

Mohamed Kanoun is a Researcher, Institut National de la Recherche Agronomique (INRA), Djelfa, Algeria.

Aziza Khalidi is a Researcher, The Collective for Research and Training on Development – Action (CRTD.A), Beirut, Lebanon.

Abdellatif Khattabi is a Professor, École Nationale Forestière d'Ingénieurs (ENFI), Salé, Morocco.

Malika Maâmri is an Agronomist, Institut National de la Recherche Agronomique (INRA), Algiers, Algeria.

Karima Maâni is a student of Gender and Development, National School for Agriculture, Meknes, Morocco.

Manar Matah is a Student/Researcher, Haut-Commissariat aux Eaux et Forêts (HCEF), Chefchaouen, Morocco.

Amel Meguellati-Kanoun is a Researcher, Institut National de la Recherche Agronomique (INRA), Djelfa, Algeria.

Abdelhakim Ouzzane is a Researcher, Institut National de la Recherche Agronomique (INRA), Algiers, Algeria.

Andrea Pape-Christiansen is a Consultant at the World Bank (temporarily located in Amman, Jordan).

Siham Rohain is a student/Researcher, École Nationale d'Agronomie (ENA), Meknes, Morocco.

Mohamed Sbai is a student/Researcher, École Nationale d'Agronomie (ENA), Meknes, Morocco.

Omar Traboulsi is a Programme Manager, The Collective for Research and Training on Development – Action (CRTD.A), Beirut, Lebanon.

Fatima Zahid is a Professeure de l'Enseignement Supérieur, Ingénierie du Développement, École Nationale d'Agriculture de Meknès (ENA), Meknes, Morocco.

Acronyms and abbreviations

AGAT	Association for the Territorial Management and Development
APW-AOC	Assemblee Populatire de Wilaya
CEDAW	Convention on the Elimination of All Forms of Discrimination against Women
CENEAP	Centre d'Études et d'Analyses d'Études et d'Analyses pour la Population et le Developpement
CERAI	Centre d'Études Rurales et Agriculture Internationale
CIHEAM	Centre International de Hautes Études Agronomiques Méditerranéennes International Centre for Advanced Mediterranean Agronomic Studies
CIRAD	Centre de Coopération Internationale en Recherche Agronomique pour le Développement, France
CNES	Conseil national économique et social, Algeria
CRASC	Centre National de Recherche en Anthropologie Sociale et Culturelle
CRTD.A	Collective for Research and Training on Development – Action, Lebanon
DFID	Department for International Development
DMN	Direction de la Météorologie Nationale, Morocco
DPAT	Direction de Planification et de l'Aménagement du Territoire, Djelfa, Algeria
DSA	Direction des services agricoles, Algeria
ENA	École Nationale d'Agriculture, Meknes, Morocco
ENFI	École Nationale Forestière d'Ingénieurs, Salé, Morocco
ENRM	Environment and natural resource management
ENSA	L'École Nationale d'Agriculture, El-Harrach, Algiers, Algeria
FADES	Fonds Arabe pour le Développement Economique et Social
FNUAP	Fond des Nations Unies pour la Population
HACCP	Hazard Analysis and Critical Control Points
HCEF	Haut-Commissariat aux Eaux et Forêts, Chefchaouen, Morocco
ICARDA	International Center for Agricultural Research in Dry Areas, Jordan
IECAH	l'Instituto de Estudios sobre Conflictos y Acción Humanitaria
IDRC	International Development Research Centre
IFAD	International Fund for Agricultural Development

ILO	International Labour Organisation
INRA	Institut National de la Recherche Agronomique, Algeria
ISO	International Organization for Standardization
ITMA	Institut Technique Moyen d'Agronomie, Djelfa, Algeria
IUCN	International Union for Conservation of Nature
MDGs	Millennium Development Goals
MEDA	MEsures D'Ajustement ou D'Accompagnement – Euro-Mediterranean Partnership
MENA	Middle East and North Africa
NARS	National Agricultural Research Systems
NGO	Non-governmental organization
RWC	Rural women's cooperatives
RGPH	Le Recensement Général de la Population et de l'Habitat, Morocco
SAGA	Social and gender analysis
SAS^2	A Guide to Collaborative Inquiry and Social Engagement (see refs)
SDC	Swiss Agency for Development and Cooperation
SMEs	Small and medium-sized enterprises
UNIDO	United Nations Industrial Development Organization
UNIFEM	United Nations Development Fund for Women (Now called UN Women)
WEDO	Women's Environment & Development Organization (WEDO)
WFC	World Fish Center
WHO	World Health Organisation

Acknowledgements

This volume presents the results of a project 'Because Gender Matters: Strengthening Social and Gender Analysis in ENRM in the Middle East and North Africa' carried out from March 2009 to March 2013. This project was coordinated by the International Center for Agricultural Research in the Dry Areas, Jordan (ICARDA) and funded by the International Development Research Centre (IDRC) of Canada. This book is part of ICARDA's and IDRC's commitment to synthesize and present relevant knowledge from their research activities that contributes to strengthening gender aspects of natural resources management.

We express our appreciation to the many people who contributed to the creation of this book and to those who expressed loudly that a gender component is lacking in natural resources management in the MENA region, compared to other regions of the world. Thanks to those who provided support, discussed and debated with us the issues presented, reviewed, offered comments, and allowed us to benefit from their experience, assisted in the editing, proofreading and design of this publication.

This participatory research project and publication would not have been possible without direct contact with and feedback from women and men in many rural communities. We dedicate this book to them as a way of expressing our thanks, with our hope the experience they shared will be of service to the young generations of our region in the Middle East and North Africa in generating a true impact on improving livelihoods.

The book was developed with contributions from many individuals. Lamia El-Fattal, former Program Officer at the Regional Office of IDRC in Cairo, followed up the initiation of the project in collaboration with ICARDA scientists through her clear vision at the project's conception. She was key in the earlier stages of the project's preparatory workshops, involving experts in a consultation workshop on strengthening regional approaches to gender in environmental and natural resources management research, and during the participatory development of the research proposal. Thanks also to Hammou Laamrani, Senior Programme Officer at IDRC Cairo, who followed the coordination and implementation of the project, providing useful insight, accompanying project coordinators in their implementation, provided his input, and constructive reflections.

Research teams from Djelfa (Algeria), Bekaa (Lebanon), Boudinar (Morocco) and Maamar (Morocco) contributed to this volume. Djelfa study: Mohamed Kanoun (Project leader), Amel Meguellati-Kanoun, Abdelmadjid Benmebarek, Malika Abdelali-Martini, Mohamed Larbi Cherfaoui, Abdelhakim Ouzzane, Malika Maamri, Mohamed Benidir, Fodil Souhila and Johann Huguenin; the Bekaa case study: Lina Abou-Habib (Project leader); Boudinar case study: Abdellatif Khattabi (Project leader), Manar Matah, Naima Fawzi, and Soumaya Ibrahim; Maamar case study: Fatima Zahid (Project leader), Asmae Harrad, Siham Rohain, Karima Maâni, Mohamed Sbai, Maria Fernandez, and Aden Aw-Hassan; project's impact: Andrea Pape-Christiansen, Malika Abdelali-Martini, and Aden Aw-Hassan; Abdellatif Khattabi, and Soumaya Ibrahim. Last but not least, our thanks go to Rima El-Khatib for her great computer skills support over the lifetime of the project.

Editorial support and facilitation in producing the final manuscript was provided by Michael Devlin and support staff of the ICARDA communications team.

Preface

Mainstreaming social and gender analysis in environment and natural resource management research in the Middle East and North Africa region

The impetus for this book came from a growing concern and unease about the unsatisfactory consideration and mainstreaming of gender in natural resource management projects in the Middle Eastern and North Africa (MENA) region as compared to other regions in the world. This deficiency is in contradiction with the wider international debate on and engagement in forums concerning the introduction of gender considerations into development and research programs in developing countries more widely.

There are a number of barriers and perceptions in the region that prevent a broader view on gender and its application in national resource management programs. Gender remains primarily considered or interchangeably looked at as only women and their roles, which blocks its natural evolvement to become transformative in development programs. In its broader definition, gender encompasses the socially constructed roles and responsibilities that are prescribed for men and women and the resulting social inequality that these impose on many aspects of development. In that context gender involves understanding the intra-household division of labor, roles, decision-making and power relationships for men, women, children and older people and the differences of these divisions in the socio-economic strata.

Our aim with this book is to initiate a new path that gives more consideration to gender mainstreaming in environmental and natural resource management activities, and to redesign approaches to gender and produce results that meet the development needs of rural communities in MENA. We explored this concept through a combination of iterative capacity development and research initiatives using a number of approaches and tools with multi-disciplinary teams who produced the outputs included in this book.

The book has evolved from an idea discussed between the Regional Office International Development Research Center (IDRC-Canada) of Cairo and the International Center for Agricultural Research in the Dry Areas (ICARDA). The driver of this idea was the weak research outputs on gender in natural resource management in the MENA region. Therefore, the main thrust to address this deficit was to enhance the capacity in the region to conduct social and gender analysis (SAGA) in environment and natural resource management research. Doing this, we feel, allows development challenges of poverty, social or gender inequality and environmental insecurity to be more effectively addressed.

From these interactions, a project was initiated whose goal was to build capacity of national research teams, and to generate innovative practices and methods that demonstrate the added-value of social and gender analysis (SAGA) methods in action research.

The generation of the outputs presented in this book resulted from a continuous capacity-building effort aimed at changing the perception and practice of the scientists involved – away from their usual way of doing research – and to consider integration of a true gender approach in their research agenda. The project has succeeded in raising awareness and building the capacity of the researchers involved. But there is still a long way to go in terms of building adequate gender analysis capacity into the diverse social environments in the MENA region.

The project sparked new thinking and approaches on several levels. The process of developing the four case studies was a unique learning experience that has set the stage for better work in the future. Through this work, a community of practice has emerged, bringing together biophysical and social science specialists with an interest in gender. These experiences have generated a strong commitment in the group that will reinvigorate the gender and natural resources management debate. We feel that having realized and faced these challenges together, researchers can create a long-term impact for women and men in rural communities.

On behalf of ICARDA, we would like to express our gratitude to the research teams, and to the gender advisors, who were part of our accompaniment to the teams. Our coordination of the synthesis and editing of material for this book would not have been successful without the contribution of each and every team member. We are confident that these shared efforts will set the stage for advancing the understanding of the motivations, implications and scope of gender mainstreaming into natural resource management, a critical factor on which the livelihoods of rural communities depend. We hope that this contribution to the body of research on gender and rural development in this region will provide a foundation on which gender research can thrive in the coming decades, and complement the many political, economic and social changes that the MENA region is experiencing today.

The Editors
Malika Abdelali-Martini and Aden Aw-Hassan

1 The emergence of the capacity development initiative on social and gender analysis in MENA

Malika Abdelali-Martini[1] *and Liz Fajber*[2]

Introduction

With increasing environmental and economic pressures, poor women and men of the MENA region are challenged to meet daily livelihood needs. These populations depend upon an increasingly fragile natural resource base for food and economic security and are increasingly vulnerable to climatic variability and change. Furthermore, poor women and men are hindered by limited access to resources and services, exacerbated by political and social inequities. Women, in particular, despite their significant role in agriculture and environment and natural resource management, own fewer and have less access to resources including land, water, livestock, inputs, credit and technologies, potentially limiting productivity (World Bank 2012).

Social hierarchies and power relations heavily influence access to and control over resources and benefits, and it is critical to understand the way institutions govern the use of natural resources. Gender-based relationships influence the division of resource-based labor and access to resources. Such relationships are dynamic and differ from one society to another, and depend greatly on wider social, cultural, political and economic processes. Therefore, a sound understanding of social differences, social inequalities and power relationships is needed to answer questions of who participates and how, and who benefits and how, from research and development interventions, projects, programs and policies (Cornwall 2003; Feldstein and Jiggins 1994; Beck and Fajber 2006).

In order for development efforts to materialize into real progress for poverty alleviation in the MENA region, analysis of social, gender, economic and institutional dimensions of environment and natural resource management including agriculture are critical in the development of appropriate interventions and policy responses. Despite significant progress made by research in tackling ENRM challenges, the effective integration of social and in particular gender analysis (SAGA) remains minimal, thereby limiting the effectiveness and impact of research, development and policy activities that seek to address poverty and environmental sustainability. Key reasons for this include a lack of capacity within the MENA region to apply SAGA in practice, limited regional resources demonstrating the effectiveness of SAGA in research, and institutional biases that neglect the importance of social and gender issues. To date, approaches for capacity development have primarily focused on building technical skills in biophysical sciences, and the institutionalization of SAGA in research, academic and development organizations of the region remains a challenge.

Therefore, in order to effectively address the development challenges of poverty, social and gender equality and environmental insecurity in the MENA region, the capacity within the region to conduct social and gender analysis in ENRM research needs to be enhanced. This challenge of capacity building, while emerging from regional experience, resonates deeply with experiences in other regions, such as in Asia (Fajber and Vernooy 2006), and opportunities exist to draw on lessons of what is effective to strengthen the institutionalization of SAGA in ENRM research.

To address this need, the International Center for Agricultural Research in Dry Areas, the International Development Research Centre of Canada and partners embarked on a regional initiative to strengthen SAGA in ENRM research activities in the MENA region. The development of this program involved initial steps to assess regional needs, critical gaps, and weaknesses in SAGA of past and on-going projects, initiatives and approaches in MENA (what is working, what is not and why); and develop a collaborative approach for an iterative capacity development program. This included review of literature, review of regional research and capacity building program and workshops with key regional experts. The following sections outline key results which emerged from these processes and discussions:

1 The importance of social and gender analysis in environment and natural resource management in the region;
2 Challenges and opportunities for social and gender analysis and capacity building in MENA;
3 The focus of the project on social and gender analysis in environment and natural resource management in MENA.

Gender, poverty and environment and natural resource management in the Middle East and North Africa: development challenges

The rapid population growth in the MENA region during the last decades is reflected in the overwhelming majority of young people in all countries. This has led to high dependency ratios, which average 54.1 in the overall region. In other words, for each 100 people of working age (15–64 years), there are 48.7 children (0-14 years) and 5.5 older people (55 years and above) (OECD 2011).

The recent social, political and economic changes in the Middle East and North Africa will certainly leave their footprints, although each country will change in its own way and at its own pace depending on its resources. This will particularly affect gender relationships while managing the available resources. In MENA as elsewhere, climate change will also impact on human mobility. By the end of the century, MENA is projected to become considerably hotter and drier. In several

MENA countries overall growth and poverty levels are strongly dependent on agriculture and agricultural related activity, and most of them depend on rain-fed agriculture, making this population highly vulnerable to reduced rainfall and shifts in rainfall timing and distribution. As a consequence important migration flows are observed (Abdelali-Martini and Hamza 2012), composed mainly of men to earn or to compensate a meagre living existence at origin where natural resource economies are insecure and highly vulnerable. Migration is also gendered.[3] The impact of male migration on women translates into additional vulnerability, through the exacerbation of existing economic disparities, increases in women's workload, disruptions of activities with gender-differentiated roles, and further limits women's access to male-dominated institutions such as extension agencies or market.

Countries in the MENA region face a number of impediments to development, including widening social and economic gaps between rich and poor, rapid population growth, and persistent gender discrimination. Poverty is actually relatively low in the Middle East and North Africa (MENA) – 17 per cent of the MENA region live below $2 a day. Poverty can be reduced by rapid economic growth (above 2 per cent per capita). However, poverty persists because it is concentrated among groups which are the last ones to benefit from growth: the illiterate or uneducated, rural dwellers, large families with many dependents. Gender inequality and a deficit of women's empowerment are increasing in the MENA region with serious implications for poverty reduction. Furthermore, gender inequality – the differential access to opportunity and security for women and girls – has become an important and visible issue for the economies of the Middle East and North Africa (Hosni *et al.* 2005, World Bank 2011). The lack of women's empowerment, through their exclusion from control of environmental assets and household income, perpetuates the status quo of women and hinders the attainment of Millenium Development Goals (MDGs) in general. Three countries in the region fall in the bottom third of countries ranked in the Gender Inequality Index (UN HDR 2011), with Morocco 104th, Sudan 128th, and Yemen at the bottom at 146th. Furthermore, equitable human development in the Arab region has been impaired by two governance deficits: one in state capacity and accountability and one in societies' empowerment (UNDP-AHDR 2009, UNDP 2011).

These negative social development trends are exacerbated by deterioration of the natural resource base (land, water, biodiversity) upon which the population, especially poor women, men and children depend. The region is plagued by high rates of environmental degradation (World Bank 2008). The most water scarce region in the world, MENA countries are challenged by frequent droughts, high climatic variability, desertification, land degradation and inefficient management systems that further stress irrigated, rainfed and pastoral agriculture-livestock systems. Worsening drought conditions compounded by increasing demand for water and a crisis in water quality are forcing the MENA countries, including the energy-rich Arabian Peninsula, to consider costly measures to confront the effects of climate change (Sowers *et al.* 2011). Water resources will become increasingly scarce with climatic change (Cabot Venton 2007), resulting in severe impacts on agricultural production systems and competition for resources that could escalate conflicts and violence at local, national, and even international levels. Urban areas also face difficulties of limited water availability and poor water quality, ineffective treatment of waste and pollution resulting in increases in disease, and the challenge of meeting food and basic needs of growing urban populations.

This deterioration of natural resources has profound impacts, particularly on poor people who depend on agricultural production systems for their livelihoods. An estimated 17 per cent of the MENA region were below $2 a day in 2005, and 2.6 million more people are projected to be in poverty because of the crisis by 2011 (World Bank 2011), and comprise the landless or families with very small marginal lands (Rathgeber 2003). For a large portion of the region, agricultural production and livestock management are central to meeting their food and livelihood needs. However, agricultural production remains limited, primarily due to fragmented and small land holdings, low and highly variable rainfall, land degradation, limited crop production and post-harvest practices, lack of inputs, and poor market linkages. Constraints to livestock production include limitation of availability and quality of pasture, forage and feed resources, disease and poor health services, marketing and pricing policies (ICARDA 2002a). The agricultural sector has been constrained by poor agricultural policies dominated by strong central government interventions, including macro-economic, trade and sectoral policies that discriminated against agriculture (ICARDA 2002b). This has contributed to a strong reliance on food imports to meet national needs for many of the countries in the region, which, with global rises in food prices, has led to severe repercussions on poor women and men who are unable to afford basic staples to feed their families.

There has been increasing interest and evidence in the last few years of the significant and essential roles of women in agricultural production, livestock management, management of natural resources, and marketing of derived products, and of the potential of investments in women and agriculture to achieve gains in productivity, food security, and livelihoods (World Development Report 2012; FAO 2011a; FAO 2011b; World Bank, FAO, and IFAD 2009; IFAD 2007a). In the MENA region, the FAO estimates that the female share of the agricultural labor force has risen from 30 per cent in 1980 to 45 per cent in 2011. However, women have less access than men to productive resources and opportunities, including land, livestock, extension, inputs, financial services such as credit and technology (World Bank 2012; FAO 2011a,b; Meinzen-Dick *et al.* 2011). Where women may have access to some material resources (for example land), they may lack the necessary inputs (for example fertilizer) or technologies or knowledge (how to effectively apply the fertilizer). This is compounded by a lack of extension services reaching women, with only 5 per cent of global extension resources directed at women; and only 15 per cent of extension personnel female[4] (FAO 1993). This lack of equitable access and opportunities for women in agriculture and ENRM imposes costs on the sector, the broader economy and society as well as on women and their families. The Food and Agriculture Organization estimates that equal access to resources for female farmers could increase yields in developing countries by 20–30 per cent, and agricultural outputs as much as 2.5 to 4 per cent (FAO 2011b). This may be even higher in other places with estimates of increased maize yields by 11 to 16 per cent in Malawi and by 17 per cent in Ghana (World Bank 2012).

Despite this evidence, policies and government agencies in the MENA region tend to reach only men[5] and neglect women's involvement. Agricultural technology development in the region has persistently disregarded the needs and priorities of women. An example of this is a project in Tunisia to introduce new winter chickpea varieties bred for resistance to cold temperature and disease that enable a longer growing season and higher yield than spring-sown varieties (El-Fattal 1996). While such traits of higher yield and cold resistance seem desirable, the variety required considerably more weeding, which is primarily done by women. Hence, adoption of this new technology was by large, wealthier farmers who could afford to hire extra labor. There was limited adoption by smallholder, poorer farmers, and where adoption did take place, this had significant implications on the workloads of women in the families. Other studies in Syria and Jordan show similar trends where women were not involved in nor consulted about technologies introduced by the project for better livelihoods (Abdelali-Martini and Al-Jawahri 2009a and Abdelali-Martini *et al.* 2009b). This policy gap, compounded by social and institutional biases, has led to inequities in women's and men's access to resources such as land, credit, inputs and training, further limiting production and poverty alleviation.

Such inadequate attention to women and men's differential access to productive resources is of particular concern where environmental degradation, limited productivity, poor incomes and increasingly global labor markets are leading to changes in the rural social fabric. The region is witnessing rapid demographic change, particularly male urban and international migration (Abdelali-Martini and Hamza 2012) for employment, which although has been documented since the late 1980s and early 1990s, is thought to be accelerating. While remittances have provided a significant input into national and family economies, this continuing migration has also had significant impacts on the social composition of communities, gendered availability of labor, management of natural resources and the structure of production in rural areas. Many rural women are now *de facto* heads of households – although the extent of which is not known – and become main resource managers without the necessary assets such as title over land, access to extension and credit systems, or voice or vehicle for participation in community resource management systems such as water users associations (IFAD 2007a; 2007b; and ICARDA 2007).

In such contexts of increasing environmental degradation and expanding global economies and markets, the vulnerabilities of poor and marginalized women and men who depend on natural resources in rural and urban areas are increasing. Research needs to tackle simultaneously the challenges of poverty and food security, while also improving the productivity and sustainability of the resource base. Such research cannot only focus on technological solutions, but needs to consider social, policy and institutional aspects that govern resource management. This includes the social and gender relationships embedded within institutions that influence access, use and decision-making about agriculture, water, and resources, affecting livelihood, food and water security.

Challenges and opportunities for social and gender analysis and capacity building in MENA

Social and gender analyis in environment and natural resource management research

While some progress has been made in addressing social and institutional aspects of ENRM research, significant gaps remain. A review of 37 projects in the region[6] (Fajber 2008) concluded that while some projects have generated information on differing roles and responsibilities of women and men in ENRM, few have tackled issues of gendered relationships and how they govern access to resources, and decision-making and control over those natural resources and derived benefits. This is in part due to a lack of understanding of gender concepts, lack of effective integration of social or gender issues in research design, a lack of capacity to implement gender analysis in *practice* and/or lack of personal or institutional commitment to follow through with this work. Despite solid social science methodologies, these are often not gendered, such as household surveys or participatory research methodologies. Even where information has been generated, researchers struggle with analysis of data through a gendered lens. However, some projects demonstrated strengths on which to build, including effective incorporation of gender and social analysis in research design and successful experiences of effective strategies to open up spaces for both women's and men's participation in research. These are outlined in Table 1.1.

This review, and an expert consultation workshop on strengthening regional approaches to SAGA in ENRM research (Feldstein and Abdelali-Martini 2008) highlighted key constraints and opportunities for integration of SAGA. Many issues moved beyond the project level to larger underlying institutional issues either within organizations or research and development systems. These include:

- weak integration of social and natural sciences;
- limited technical and analytical capacity in the region;
- absence of strong, documented case studies to convince decision-makers of the added value of gender and social analysis in research;
- difficulty in reaching women and men in marginal environments;
- insufficient attention to gender in agriculture and NRM policies;
- weak systems of accountability; and
- inadequate knowledge management network for peer support and learning.

However, there were also a number of opportunities on which to build. These included, for example, increasing attention to gender in ENRM research and development (e.g. in the Arab Human Development Report UNDP-AHDR 2009, and now more recently by FAO and World Bank); the availability of gender research questions, checklists and audit tools[7]; networking of researchers working on SAGA in ENRM for exchanges of approaches and methods; promotion of

Table 1.1 Weaknesses and strengths in research approaches to gender in ENRM

	Weaknesses	*Strengths on which to build*	*Approaches*
Conceptualizing gender	Range in understanding of SAGA, and confusion that gender = women Discomfort of ENRM researchers in engaging in social and gender equity issues View of gender as being donor led, external	Improved understanding of concepts, but challenges of implementing in practice Donor emphasis provides platform and budgets to engage in gender issues	Training and support during field-based practice Stronger M&E enforced by donors to document approaches
Research methods	Research methods frequently do not include a gender component, such as household surveys, PRA methods Lack of effective consideration of SAGA, especially gender, in research design Limited gendered analysis of data Limited monitoring, evaluation and accountability Research on gender mostly limited to diagnostics, roles and responsibilities	Solid rigorous social science methodologies in projects that can be engendered (e.g. sex-disaggregated studies, qualitative methods) Strong efforts towards inclusive participatory tools, livelihood approach Integration of SAGA in research questions, hypothesis Project evaluations can stimulate integration of SAGA Including women in activities, move beyond to empowering and transformative approaches	Engender methodologies with balance between qualitative and quantitative methods Build on impact assessments to measure how projects impact on the lives and empowerment of women and men of different social groups Strengthen research on social and power relations that influence institutions at multiple levels (household, community, policy) that govern resource access Communicate results of research to decision-makers

Key: M&E = monitoring and evaluation; PRA = participatory rural appraisal success stories such as women's projects that have improved their economic standing through production of oil from oil seeds and the planting of trees; building on existence and growth of women's machineries in many countries. More recently, and subsequent to this program starting, new tools and indicators have been developed such as the Women's Agricultural Empowerment Index. Furthermore, as Meinzen-Dick *et al.* (2010) elaborate, it is key not only to focus on the production and implementation efforts, but also to integrate gender at all points of the agriculture and ENRM innovation cycle: priority setting, conducting research and development, design and implementation of extension systems, and evaluation and impact assessment.

Capacity building for social and gender analyis in environment and natural resource management research

In the design of this program, capacity building was identified as an integral part of research or institutionalization, not separate from research implementation.

Experiences demonstrate that short-term training programs and courses are useful to improve exposure to SAGA concepts and methods, but such training has limited impact if it is not complemented with field-based technical support (Fajber and Vernooy 2006; Fajber 2008; Feldstein and Abdelali-Martini 2008). Researchers engaging in any capacity building program need time for 'learning by doing' through iterative learning, field-based practice and ongoing mentoring. Priorities for research need to be developed with farmers and local people. Hence, capacity building efforts must be well grounded in research activities. Regional relevant resources are needed such as regional case studies, training materials and research tools in Arabic (for example checklists, gender issues to consider in ENRM). Moreover, long-term investments in individuals and organizations are needed to build a regional pool of expertise and strengthen peer exchange and learning.

Key lessons for capacity strengthening from MENA and Asia:

- limitation of short-term training, lectures;
- grounding capacity strengthening in research, 'learning by doing';
- ongoing mentoring;
- opportunities for peer review, networking and exchange;
- value of local cases, resources in local languages;
- importance of institutional support to implement SAGA;
- investing in committed scientists over the long term.

Social and gender analysis research and the capacity building project

Building on the lessons learned above, ICARDA and IDRC hosted a number of regional experts and partners to collaborate to develop a program 'Because Gender

Matters: Strengthening Social and Gender Analysis in ENRM in the Middle East and North Africa'.

The project employed a multifaceted strategy that involved four key components: strengthening research, capacity building, organizational support and communicating results.

1 **Rigorous applied and transformative research** that integrates SAGA effectively to address challenges of poverty and social/gender inequity, while also contributing to the strengthening of food, water and environmental security. The project identified four applied research projects.
2 **Capacity strengthening** researchers to implement action-oriented research on SAGA. Efforts were designed to be iterative, long-term and grounded in field-based research activities. The project aimed to provide technical support to deepen current research approaches, methodologies and analysis, and improve peer learning and networking among researchers in the region, and development of regional resource materials in Arabic.
3 **Organizational and institutional environments,** with consideration of how these environments (for example norms, policies, support from senior managers) in which researchers work may facilitate or hinder the integration of SAGA. The project enabled structured discussions of challenges, successes and strategies with researchers and other stakeholders.
4 **Communicating results, methods and the added value of SAGA in ENRM research to different users.** The program developed different strategies to communicate results and lessons to different users (policy makers, researchers, civil society). Case studies to demonstrate effectiveness and production of diverse media and materials are critical.

The program aimed to generate new regional knowledge and empirical evidence to demonstrate the value added through SAGA to ENRM research; improved understanding of key issues in SAGA and ENRM by researchers, policy makers and civil society actors in MENA. MENA researchers with deepened practical knowledge and capacities to integrate SAGA in research approaches and design in ENRM related projects; improved methodologies and relevant resource materials for integrating SAGA in ENRM in MENA; and an informal network of researchers, civil society organizations, and policy makers engaged in SAGA and ENRM in the MENA region have strengthened their interaction, learning and exchange.

Overarching research questions

Some of the research identified two main concerns: (1) the application of SAGA to address poverty and ENRM development challenges; and (2) the integration of SAGA in projects and programs.

The issues and questions on SAGA, poverty and ENRM include:

Access to and use of natural resources

How do women and men perceive differential access to, decision-making regarding and control of environment and natural resources? How do gendered institutions, norms and culture govern this access and decision-making and how is this changing in contexts of environmental change, economic stresses (markets) and conflicts?

Technologies and markets

How can technical processes (development, delivery and access to tools, technologies) be made more gender-sensitive in the context of climate change? How can gendered market access for poor rural women and men be enhanced?

Vulnerability and adaptation

What are gendered determinants of women, men and children's vulnerability to environmental risks and economic stresses? What are gendered responses and adaptation strategies to these environmental risks and economic stresses?

Empowerment

How can rural women and women's organizations (formal and informal) be empowered in the context of sustainable natural resource management (for example improving access to markets)? How can these women's organizations be supported to challenge, change and influence inequitable institutions at household, community, or state levels?

The issues and questions on the integration of SAGA in projects and programs include: (1) what are existing constraints (policies, institutional, organizational) that hinder the inclusion of SAGA? (2) How can these best be addressed? (3) What SAGA related indicators and tools are needed to evaluate success? (4) How can results and lessons, including the added value of SAGA in ENRM research, best be communicated and demonstrated?

Program design: case studies and capacity building efforts

To select case studies, a competitive call for proposals was launched in April 2009 inviting research proposals focusing mainly on gender and livestock production and water tenure and management systems. However, it also opened the door for other themes such as changing rural social landscape, property rights and land tenure, engendering markets, vulnerability to climate change, impacts of globalization, and technology development.

A technical group[8] reviewed the eight proposals submitted, and four case studies were selected. Key criteria for selection of the proposals included (among others): potential to demonstrate the added value and effectiveness of implementing SAGA research in an ENRM context (this may include the potential to enhance the uptake and maximize the impact of NRM interventions); potential to enhance science quality in SAGA (including new methodologies); addresses research problématique and a sub-set of overarching research questions; location of lead research institutes, research and beneficiaries are in MENA region; and strong commitments by applicants and their host organizations to participate not only in the overall program, but to the sustainable objective of strengthening SAGA in ENRM research. Members of the technical group provided inputs and support to strengthen and refine the proposals, and ongoing support was provided to the researchers by the ICARDA team and resource persons.

The four selected case studies were (original titles as submitted):

1 **Algeria:** Options for enhancing the utilization of local knowledge on livestock production, livelihood improvement and gender in Djelfa governorate.
2 **Lebanon:** Factors preventing women's rural cooperatives from accessing new external and internal markets: a gender analysis perspective.
3 **Morocco, Boudinar:** Gender analysis of the effects of declining water resources, resulting from climate change, on rural communities and their coping mechanisms in Boudinar, Province of Driouech.
4 **Morocco, Maâmar:** Gendered valuation of local knowledge for enhancing the utilization and conservation of aromatic and medicinal plants in the Atlas mountains of Morocco; the case of Douar Maâmar.

The program was designed to strengthen capacity using the following modalities such as a true engagement for capacity building over the entire project life where training courses were provided in addition to continuous mentoring and peer learning and exchange. This was also supported through field-based support for 'learning by doing' in applying skills, tools (Feldstein, H.S. and Jiggins 1994) and methods in the four research projects. More details on the capacity building experience of the program are outlined in the last chapter of this volume.

The strong commitment of researchers to partner in this pioneering initiative constituted the seeds for our collaborative journey of research through capacity building in SAGA through this project. It is hoped that through wide communication of research results and lessons of this effort, researchers, research managers, donors, policy-makers and civil society will have a stronger understanding of the critical importance and added value of SAGA research. This will help contribute to a larger goal of institutionalization of SAGA in ENRM research and development activities.

Each of the case studies is outlined in the following chapters. Some cases focused more on generating evidence and descriptive understanding of the different and complementary gender roles and related knowledge in agriculture and natural resource management in the region; factors that may constrain or inhibit women from accessing productivity or economic gains from agriculture and NRM (Lebanon); strategies for coping with or adapting to environment and economic shocks (Algeria, Morocco-Boudinar); and action-oriented approaches to improve livelihoods of women through addressing these constraints (Algeria, Morocco-Maamar). Despite capacities and institutional constraints, the efforts of the overall program have succeeded in being action-oriented and transformative.

The program sought to address organizational and institutional constraints to SAGA in ENRM, such as lack of institutional or organizational commitment and engagement of senior managers. Though addressing the institutional elements was in the original design of the program, this was an area that proved challenging as specific activities were not defined nor resources allocated. However, some strategies included meetings with senior managers at national institutions to raise the status of the SAGA study and gain their participation in the final workshop; lectures on SAGA to students at some of the institutions; and engagement of post-graduate students in the program as part of their thesis work and results dissemination to a wide audience of researchers, students, decision-makers, international observers, experts in the field and media.

The last two chapters of this book outline in more detail the experience and lessons learned from this overall program. These focus on 1) lessons from the integration of SAGA in ENRM research, particularly through using participatory approaches, and 2) reflections on the capacity development experience of the project, impacts and lessons learned.

Concluding remarks

Enhancing scientists' gender research skills is one of the most important capacity building investments that an organization can make. With strengthened capacity through social and gender analytical skills, women and men scientists, managers and leaders can better tackle poverty alleviation through the identification of differential concerns and constraints of the community members, promote better policy advocacy, build collaborative relationships and diverse partnerships, better serve their organizations' interests, and manage the overall impact and change more effectively.

The SAGA project is unique for a number of reasons: it was composed of and owned by a multidisciplinary team of women and men scientists, students, and other practitioners from the MENA region; it provided a unique opportunity to build capacity in social and gender research among biophysical scientists working on water management, agro-biodiversity, rangeland and livestock scientists, in addition to sociologists and rural development specialists. Mentors and resource persons from International Center for Agricultural Research in the Dry Areas and technical experts provided support to regional researchers to analyze social and gender dimensions of ENRM specific to the MENA region, and identified potential solutions to address inequalities and production and economic constraints. These projects tackled specific challenges that women and men face in order to secure decent livelihoods and wellbeing for their families, while exploring ways to sustain natural resources through effective management. This program constitutes an initial step towards strengthening efforts to address social and gender dimensions of ENRM. However, as explained above, such efforts need to be long-term, iterative, and have strong support to address the institutional environment. We need to build on successful efforts and galvanize current increasing attention to the role of women in agriculture, natural resource management and food security. Much more remains to be done.

Notes

1 International Center for Agricultural Research in the Dry Areas (ICARDA), Amman.
2 Work related to this chapter was conducted in personal capacity, when working as an independent consultant prior to joining Department for International Development (DFID).
3 In fact the majority of migrants are men, but there are also a few women. In our study 14 per cent of the total migrants were women.
4 These statistics are the last available statistics from the FAO, with a survey of 97 countries in 1988-89, and estimates of women personnel in 1993.
5 Policies and programs may not even reach men or women in marginal and more isolated areas (IFAD 2007b).
6 This review also included a review of key relevant literature from the region (in English) and interviews with research and donor agencies.
7 Patricia Morris, The Gender Audit (2003) InterAction, Washington, DC: A paper reviewing the use of the gender audit with NGOs in Africa is Meryl James-Sebro *et al.* (2012) Revealing the Power of Gender Mainstreaming, Enhancing Development Effectiveness of Non-governmental Organizations in Africa. These two publications plus one from Asia can be purchased from Interaction atwww.interaction.org. A gender audit has also been conducted at one Internaiional agricultural research center in Africa, the International Livestock Research Institute (ILRI). The ILRI Report can be found at www.ilri.org.
8 Malika Abdelali-Martini (M.Martini@cgiar.org); Aden Aw-Hassan (A.Aw-Hassan@cgiar.org), Lamia El-Fattal (lelfattal@idrc.org.eg); Annina Lubbock, a.lubbock@ifad.org; Monica Kapiriri (mkapiriri@yahoo.co.uk); Ruth Meinzen-Dick (R.Meinzen-Dick@cgiar.org); Adriana Bruggeman (A.Bruggeman@cgiar.org); Barbara Rischkowsky (B.Rischkowsky@cgiar.org); Alia Gana (alia_gana@yahoo.fr); and Nicholas Pasiecznik (N.Pasiecznik@cgiar.org).

References

Abdelali-Martini, M., Amri, A., Ajlouni, M., Assi, R., Sbieh, Y. and Khnifes, A. (2008) 'Gender dimension in the conservation and sustainable use of agro-biodiversity in West Asia', *Journal of Socio-Economics* 37(1): 19.

Abdelali-Martini, M. and Al-Jawahri, N. (2009a) 'How much dairy processing contributes into gender livelihoods in Tafileh and Showbak communities in Jordan', Socioeconomic component of the M&M III project. Research Report ICARDA and NCARE January 2009: Amman, Jordan. Unpublished report.

Abdelali-Martini, M. and Hamza, R. (2012) 'How do migration remittances affect rural livelihoods in drylands?', *Journal of International Development*, online at wileyonlinelibrary.com, DOI: 10.1002/jid.2896.

Abdelali-Martini M, Salahieh H., and Hamed F. (2009b) 'How much dairy processing contributes into gender livelihoods of Hama province: the case of Sheikh Hilal and Abu Al-Fashafish communities in Syria,' Socio-economic component of the M&M III project. Research report ICARDA and GCSAR: Douma, Syria.

Beck, T. and Fajber, E. (2006) 'Natural resource management, poverty, inequality and gender in Asia', in S. Tyler (ed.) *CBNRM: Action Research and Policy Change*, London: ITDG; Ottawa: IDRC Books.

Cabot Venton, C. (2007) *ERM for WaterAid: Environmental Resources Management: Climate Change and Water Resources*, London: Wateraid.

Cornwall, A. (2003) 'Whose voices? Whose choices? Reflections on gender and participatory development', *World Development* 31(8): 1325–42.

El-Fattal, L. (1996) Women in agriculture in West Asia and North Africa: a review of literature. Consultative Group for International Agricultural Research (CGIAR) Gender Program Working Paper No. 10. Washington, DC: CGIAR Secretariat, World Bank.

Fajber, E. (2008) 'Integrating gender analysis in environment and natural resource management research in the Middle East and North Africa: a review of challenges and strategies'. Unpublished report for IDRC and ICARDA.

Fajber, E. and Vernooy, R. (2006) 'From voice to choice: lessons from a regional capacity development initiative to strengthen social and gender analysis in participatory natural resource management research', *International Journal of Agricultural Sustainability* 4(2): 143–53.

FAO (1993) *Agricultural Extension and Women farm Workers in the 1980s*. Rome: Food and Agriculture Organization of the United Nations.

FAO (2011a) The role of women in agriculture. ESA Working Paper no. 11, 2 March. Prepared by the SOFA team and Cheryl Doss, Rome: Food and Agriculture Organization.

FAO (2011b) 'The state of food and agriculture 2010–11: women in agriculture – closing the gender gap for development'. Rome: Food and Agriculture Organization.

Feldstein, H. and Abdelali-Martini, M. (2008) 'Strengthening regional approaches to SAGA in ENRM in MENA: expert consultation workshop 9–10 April. Aleppo, Syria. Summary Report.

Feldstein, H.S. and Jiggins, J. (eds) (1994) *Tools for the Field: Methodologies Handbook for Gender Analysis in Agriculture.* West Hartford, CN: Kumarian Press.

Hosni, D., Sandberg, M. and Chanmalla, A. (2005) 'Inequality among women and its impact on economic growth: the case of MENA in sustainable human development in the human century, Vol. II', online at www.eolss.net/Sample-Chapters/C11/E6-60-05-02.pdf

ICARDA (2007) *Improving Livelihoods in Dry Areas: ICARDA Strategic Plan 2007–2016*, Aleppo: ICARDA.

ICARDA (2002a) 'Improving the livelihoods of rural communities in the Nile Valley and Red Sea region: sustainable crop, livestock, land and water management', Improving the livelihoods of rural communities in the Nile Valley and Red Sea region project concept note.

ICARDA (2002b) 'Developing sustainable livelihoods of agropastoral communities of West Asia and North Africa', Mashreq-Maghreb project phase III concept note.

IFAD (2007a) 'Looking for new directions', Gender mainstreaming and women's empowerment workshop Report, 16–17 May, Rome: IFAD.

IFAD (2007b) *IFAD Strategic Framework*, Rome: IFAD.

IFAD (2005) 'IFAD's strategy for an equitable development for women and men in the Near East and North Africa Region.', Rome: IFAD.

Meinzen-Dick, R., Johnson, N., Quisumbing, A., Njuki, J., Behrman, J., Rubin, D., D. Peterman, D. and Waitanji, E. (2011) 'Gender, assets and agricultural development programs: a conceptual framework', CAPRi Working Paper no. 99, International Food Policy Research Institute: Washington, DC: Online at http://dx.doi.org/10.2499/CAPRiWP99

Meinzen-Dick, R., Quisumbing, A., Behrman, J., Biermayr-Jenzano, P., Wilde, V., Noordeloos, M., Ragasa, C. and Beintema, N. (2010) 'Engendering agricultural research', IFPRI Discussion Paper 00973, Environment and Production Technology Division Washington, DC: IFPRI.

OECD (2011) *Socio-Economic Context and Impact of the 2011 Events in the Middle East and North Africa Region MENA-OECD Investment Programme*, online at www.oecd.org/mena/investment/49171115.pdf

Rathgeber, E. (2003) *The Gender Relations of Agricultural Production in the MENA Region: Some Preliminary Thoughts*. Cairo, IDRC Office for the Middle East and North Africa Region.

Rathgeber, E. (2002) 'Female and male CGIAR scientists in comparative perspective', Working paper no. 37. CGIAR Gender and Diversity program, Research. Online at http://library.cgiar.org/bitstream/handle/10947/2739/37_Female%20and%20Male%20CGIAR%20Scientists%20in%20Comparative%20Perspective_genderdiversity_WP.pdf?sequence=1

Rathgeber, E.M. (1990) 'WID, WAD, GAD: trends in research and practice', *The Journal of Development Areas* 24: 489–502.

Sowers, J., Vengosh, A. and Weinthal, E. (2011) 'Climate change, water resources and the politics of adaptation in the Middle East and North Africa'. *Climatic Change* 104: 599–627. DOI 10.1007/s10584-010-9835-4.

UNDP-AHDR (2009) *Arab Human Development Report: Challenges to Human Security in the Arab Countries*. UNDP.UNDP (2011) *Arab Development Challenges Report 2011: Towards the Developmental State in the Arab Region*. Online at: http://arabstates.undp.org

UN HDR (2007) *Human Development Report 2007/2008. Fighting Climate Change: Human Solidarity in a Divided World*. Online at: http://hdr.undp.org/en/media/HDR_20072008_EN_Complete.pdf

World Bank (2011) *Gender and Development in the Middle East and North Africa*, online at http://siteresources.worldbank.org/INTMENA/Publications/20262206/genderoverview.pdf

World Bank (2011) *Poverty in the Middle East and North Africa*, online at http://siteresources.worldbank.org/INTMNAREGTOPPOVRED/Resources/MNA_Poverty_EN_Final.pdf

World Bank (2012) *World Development Report 2012: Gender Equality and Development*, Washington, DC: The International Bank for Reconstruction and Development/The World Bank.

World Bank, FAO and IFAD (2009) *The Gender in Agriculture Sourcebook*. Washington, DC: The International Bank for Reconstruction and Development/The World Bank.

World Bank (2008) *Environment in MENA: Sector Brief*, Washington, DC: The International Bank for Reconstruction and Development/The World Bank.

2 Marginalisation of women's work as a result of changes in pastoral societies and natural resources

Mohamed Kanoun[1], Amel Meguellati-Kanoun[1], Malika Abdelali-Martini[2], Johann Huguenin[3], Mohamed Larbi Cherfaoui[1], Abdelhakim Ouzzane[1], Abdelmadjid Benmebarek[4], Malika Maamri[1] and Souhila Fodil[5]

Introduction

Livestock rearing is of high importance for both the populations in the steppe regions and for the national economy. Indeed, the animal farming sector contributed 51 per cent of the agricultural GDP and this was probably higher if other intermediate values in animal farming were evaluated more appropriately (Akhilu 2002). This sector also provides raw materials such as milk, wool and camel hair required for traditional economic activities.

The changes that have taken place over recent decades in pastoral societies have led to socio-economic changes that have had repercussions on the organisation of work. Women who conform to traditions and customs have been heavily impacted by these changes (Boukhobza 1982. The settling of nomadic and transhumant households, the increasing use of feed supplements (mainly barley and bran) as feeds (moving away from pastoral feeding practices based solely on pasture) and the private acquisition of rangelands constitute the main developments in the steppe (Moulai 2008).

These changes have resulted in severe deterioration of the natural resources of the rangeland, the basis for agro-pastoral production systems in Algeria. The deterioration of plant cover in the steppe has been confirmed by ecological and agro-pastoral analyses (Le Houérou 1995; Nedjraoui 2003; and Tarhouni *et al.* 2007).

This deterioration, linked to the changes in animal production systems, has now reached the point that it is threatening the viability of traditional income-generating craft activities, especially those related to the production of a wide range of *qashabiya*. Sheep wool and camel hair are the main raw materials. These craft activities hold social significance as they involve women from pastoral and agro-pastoral societies.

We studied six agro-pastoral communities where raising ruminants is the main source of income, and analyzed the loss of craft know-how in pastoral societies. Our main research question was: how have the deterioration of productive natural resources and the changes in animal production systems contributed to the availability of raw material for *qashabiya* production?

Our work was centred on the following three main hypotheses: (1) The deterioration of pastoral resources, combined with adverse effects of climate change have led to social and technical reorganisations in the economics of *qashabiya* production mainly in terms of the availability of related raw material. (2) The changes in animal-raising practices and the access to fodder have increased the women and men's workload. (3) The increased distance of the production centres from the raw materials (especially camel hair) have resulted in scarcity of raw material for women.

Study location

The Wilaya of Djelfa, capital of the steppe, is located in the middle of the Algerian steppe, (Figure 2.1). Around 85 per cent of the total area, estimated at 3,225,635ha is rangelands, with a total of about 2.9 million sheep. Three main methods of animal farming are practised by livestock producers: transhumance, semi-transhumance and sedentary farming. Animal-farming activities are the main sources of income for the majority of households (Kanoun 1996). The population of the Wilaya is 1,119,542 inhabitants (DPAT 2009), currently divided among 36 districts and grouped into 12 sub-districts. In terms of demographics, 60 per cent of the population is under 24 years old (DPAT 2010).

Figure 2.1 Location of the study area: the Wilaya of Djelfa

Role of tribes

The population of the Wilaya of Djelfa belongs to three large tribes: the Ouled Nail tribe controls two-thirds of the area of the three communities (Annex 1). The remaining third of the area is split between the El Abaziz, the Essahari and the Ouled Rahman. Each tribe is divided into several fractions[6], each with its own area which is recognised and respected by all of the pastoral communities. However, it should be noted that since the 1980s, the decrease in joint ownership of rangeland and the development of individualism have upset the tribal order.[7] Due to this, former pastoral land has come to be used for other farming activities of unknown sustainability, breaking up the area into smaller patches, sealing off transhumance corridors and cutting back the amount of the most productive rangeland. Before the agrarian reforms affecting joint ownership, the 'Arch' body or tribe administered the rangeland, managed it and decided the routes taken, camping locations and grazing durations.

Role of women

All women have roles and activities connected to animal farming, regardless of the farming method. They are involved mainly in feeding young animals, watching over animals close to the place of residence, etc. How closely they are involved differs between systems. However, they do not have the right to participate in the management of the rangeland, even in connection to the livestock. In these patriarchal societies, the women are excluded from the management of the rangeland. Even if a woman has land, it is used and managed by a male member of her family (father, brother, husband or son). Currently, the implementation of the 2000 national agricultural development plan (PNDA – Programme National de Développement Agricole) which provides special measures for the modernisation of farming has not allowed women to benefit to the same level as men.

Climatic situation

The climate of the Djelfa region is characterised by a dry season that lasts five months or more. Precipitation is very irregular and the seasonal distribution greatly varies from one year to another. On average over the 2000–2010 period, precipitation did not exceed 260mm per year. The steppe regions suffered several droughts during that period: three consecutive years were severe and classified as agricultural disasters (2000, 2001 and 2002). This natural phenomenon has greatly magnified pastoral land degradation. In the past, steppe rangeland was able to provide all of the livestock feed requirements. Undoubtedly, the animal husbandry system was more mobile and the complementarity between the regions allowed animal feed requirements to be met. Currently, rangeland provides on average only 30 per cent of livestock feed requirements in the region (Kanoun *et al.* 2007). The rest is provided by cultivated resources and purchased at high prices. In order to reduce exposure to fluctuations in natural resources, pastoral societies have made major changes to their production systems (Hatfield and Davies 2006).

Methodology

In order to answer the research questions, the method adopted was based on a systemic approach, stressing the gender aspect by referring to male–female social relationships, structural differences characterising these in terms of their socially assigned and culturally justified roles, status and function (Damamme 2005). In our study, gender analysis allowed us to shed light on the implications, issues and problems faced by women and men, due to the deterioration of productive natural resources which supports the livestock production thus securing the raw materials for the production of *qashabiya*. The system under study is complex with variable environment, plants and animals, used by people for different purposes, and under noticeable changing climate conditions. This requires inter-disciplinary effort involving social sciences and biophysical sciences (Bako-Arifari and Le Meur 2001).

Specifc research tools

We used the tools from two participatory approaches: the Accelerated Participatory Research Method (MARP – Méthode Accélérée de Recherche Participative) and the Social Analysis Systems (SAS[2] – systèmes d'analyse sociale). The use of these qualitative tools is justified, on the one hand, by ease that they are adapted to the context of this study marked by sociocultural and economic changes and, on the other, by the fact that their use allows us to include the perspectives of rural women and men. These tools were carefully selected following our observations of agro-pastoral family production units and individuals (five people with different statuses: livestock producers, craft workers, traders, etc.) and groups (two groups of livestock producers and craft workers and a group of craftswomen) and interviews with key local people with a good knowledge of the ecological, sociocultural and economic history of the region. The selection of people and family production units was random and the questions mainly concerned processes of animal husbandry, use of natural resources and expertise on livestock products, especially those using camel hair This pragmatic approach allowed the identification of key contact people (women and men) to access information in an environment where traditions have an important impact on gender relationships.

The following tools were used to confirm or reject our hypothesis

The historical profile combined with the timeline: the historical profile was used to analyse the land use patterns, the natural-resource dynamics and to explore the relationships between communities and their natural environments (Vabi and Mukong 2002). We gathered the information from a group of shepherds and nomads aged 65 to 85 years (this work was a reconstruction of a retrospective memory from the respondents). We then used the 'timeline' in a meeting that lasted about 90 minutes. This was critical to understand the events that had contributed to the deterioration of natural resources, and contributed to understanding the changes in the roles of men and women over time.

Scoring and prioritisation grids combined with the scoring method: this tool allowed us to classify and prioritise the events that had the greatest impact on the deterioration of natural resources according to the perception of livestock producers. According to Chevalier *et al.* (2008) the use of this tool allows sensitive information to be obtained. Furthermore, we carried out an evaluation of the repeated effects of every event contributing to the deterioration of the natural resources. Listing the events alone was insufficient for appreciating their negative impact as several events were mentioned several times. To this end, we deemed it appropriate to score and classify the repeated effects of drought and highlight the links to other events.

Resource maps: in the absence of cadastral information, this tool nonetheless allowed the interpretation of information and land observations, and the illustration of arguments relating to the changes to land use patterns. A group interview involving an heterogeneous group of people consisting of small ruminant producers (of different ages) revealed major changes that had affected natural resources. According to Wilde and Aia Vainio (1995), resource maps are useful to identify preferred species used by the women or the men, or by both.

Satellite images and GIS technology: the relationship between land use patterns and ecosystem dynamics is central in our research. Remote sensing and GIS strengthened and improved our analysis about the changes in the forages at the study site by providing more accurate information (Mottet 2005).

Finally, the changes impacting local populations were approached through participatory group discussions and the resulting information was then checked with key informants at local institutions.

Furthermore, several meetings were conducted with livestock producers, craft workers, women and youth. We aimed at collecting information on how the changes to livestock production systems affected gender relations and the supply of raw material needed to make the *qashabiya*. Paying attention to gender allowed the specific circumstances of each person to be taken into account, in relation to the various problems that have affected the '*qashabiya* sector'. As such, four groups were formed to collect information differentiated by gender.

The first group consisted of around 10 livestock producers aged between 50 and 70 years. The meeting took place at the farm, and we used the problem tree to seek information for the last two research questions. For this, the problem tree (Chevalier *et al.* 2008) was the most suitable tool for providing answers to these hypotheses. Indeed, the use of this tool revealed the main causes for the change and how the supply of raw material needed to make the *qashabiya* has changed. The changes to small ruminant production systems led to changes in herding practices, especially those related to animal feeding. The researcher-led meeting lasted around 90 minutes.

In order to understand the changes to household member activities, the greater distance from the production centres and the availability of raw material (wool and camel hair), we organised three meetings, two of which were with women living in the steppes and one with a group of men camel producers. Here, we used three tools.

The problem tree: As the main area where most of the production and marketing of *qashabiya* are made, Messaâd was selected for this investigation in collaboration with a resource person (a woman) working for a women's association. This facilitated the access to this category of crafts and the discussions were rich especially because they took place at the craft workshops. This meeting, led by a female researcher in collaboration with a resource person (president of a women's association), lasted around 120 minutes. It unveiled the main causes of natural resource degradation as the exploitation of the rangelands for cereals and the appropriation of parts of the steppe, the effects this has on livestock production and consequently negative effect on raw material supply from sheep and camels.

The activities' diagram was used in the Deldoul area and included women from an agro-pastoral community where animal farming is the main source of income. The information collected was strengthened by field-study observations and showed the important contributions of women to farming and domestic activities. Led by a female researcher, this meeting lasted 90 minutes.

Force field: according to Chevalier *et al.* 2008, the use of this tool helps in understanding different viewpoints about the factors perpetuating the problem versus those mitigating it ; and those that are preventing it from becoming worse. This tool was used with a group of camel producers from the south of the Wilaya (governorate) of Djelfa. It identified the causes of low production of raw material needed for *qashabiya* production. The meeting was led by a researcher who could speak the local dialect. It is important to highlight that being able to speak the local dialect is a major advantage in this type of participatory research. The meeting lasted 60 minutes.

Results

Scarcity of natural resources and of agro-pastoralism: key factors determining the supply of raw material needed to make the *qashabiya.*

Using the timeline and historical profile we identified, on one hand, the events that contributed to the deterioration of natural resources and, on the other hand, the dynamics of the process over time and place. For livestock keepers in the steppe, the droughts of 1970–1971 constituted the main driver for the deterioration of natural resources. During these difficult periods, the state would often make emergency interventions with 'drought plans' such as supply of subsidised feed (barley and fodder) to fill the feed shortage. Our interviews with livestock producers revealed the supply of subsidised barley as a major turning point in the livestock producers' perceptions on the importance of cereal production and the growth of the fodder market in parallel. Today, all of these factors explain the ability of the livestock producers to live in a fragile area that has scarce pastoral and natural resources. The results obtained using the historical approach carried out with a group of elderly livestock producers showed the almost complete disappearance of the best pastoral species such as esparto grass (Stipa tenacissima), white wormwood (Artemisia herba alba) and albardine (Lygeum spartum). They reported that this deterioration process was intense during the 1990s. The livestock producers highlighted that certain tribe members, in connection with few local authorities within the Wilaya of Djelfa, had taken advantage of this period of insecurity to plough the best natural rangeland.

On a scale ranging from 1 to 9, this natural factor was rated 9 (1 = lowest, and 9 = highest) (Table 2.1). In addition, scoring and prioritisation grids were used to analyse this information, and attributed a high score for the repeated effect of droughts, as indicated by a high score of 13 (Table 2.2). Indeed, the livestock producers indicated that there had been several droughts in recent decades in the steppe.[8]

Moreover, satellite images and the use of GIS confirmed our results from the various participatory surveys. Rangelands in El-Guedid decreased by 27 per cent over 30 years (Table 2.3) and sandy and cultivated areas increased by 12 per cent and 11 per cent respectively (Table 2.3).

Pastoral areas are likely to be no longer exclusively devoted to livestock producers' activities, despite the importance of the farming of small ruminants (Figure 2.2). The introduction of mechanisation and tractors has encouraged livestock producers to invest in dry and irrigation farming. The scale of cereal production in rangelands is increasing with no consideration shown to the preservation of natural resources. Results of the various semi-structured individual and group interviews showed that livestock producers are growing rainfed cereals without any inputs, as a measure to mark out their territory and take over rangelands through the practice of illegal appropriation 'gdel'.

Today, this phenomenon is the origin of illegal privatisation[9] of shared land and the development of individualism, which promotes the unsustainable use of natural rangeland. The analysis revealed that the lifting of joint ownership of steppe land resulted in informal fragmentation of the land and division of plots between heirs has reduced the per capita (illegally) cultivated area. With the aim of compensating for this reduction in cereal production, livestock producers and tribes[10] that do not have flocks prepare and plough rangelands for renting out[11] to owners of large herds.[12]

Because of the deterioration of pastoral resources, cereal production is becoming an unavoidable component of steppe sheep production systems as they are used in several forms: green barley, grain, crop stubble and straw. However, our results confirm the extent of the deterioration of rangelands caused by 'Gdel'. Regarding the complexity of the issue, livestock producers gave 'forbidding Gdel' in rangeland areas a score of 9, as an important solution for reducing the pressure on and the deterioration of productive natural resources.

According to the statements made by the livestock producers, pastoralism based exclusively on pastoral resources is currently rare, practised only by a minority of people, mainly the landless. Even so, the analysis of secondary data shows that the decline of pastoralism has not corresponded with a decline in animal farming. Quite the opposite, it has been growing constantly: the overall sheep population in the country rose from 10 million during the mid-1970s (1975–1977) to over 19 million in 2009–2011.

Table 2.1 Scoring the significance of events on the devaluing of local knowledge according to participants' perceptions

Event	*Drought*	*Introduction of the tractor*	*Creation of pastoral cooperative*	*Subsidy of livestock food*	*APFA*	*Audio-visual and communication technologies*	*Food and textile industry*	*PNDA*	*Training centres*
Scoring	9	6	4	6	4	8	7	5	9
Prioritisation	9	5	1	4	2	7	6	3	8

Key: APFA = Accession à la propriété foncière agricole (APFA: Access to agricultural land) ; PNDA: Programme National de Développement Agricole (NADP: National Agricultural Development Programme).

Source: INRA-SAGA 2010: our fieldwork investigations.

Table 2.2 Scoring the significance according to the frequency of the event

Event	*Drought*	*Introduction of the tractor*	*Pastoral cooperative*	*Subsidy of livestock feed*	*APFA*	*Audio-visual ICT technologies (TV)*	*Food and textile industry*	*PNDA*	*Training centres*
1960			+						
1970	+++	+					+		+
1980	+++	++		+	+		+		++
1990	+++	+++		++	+	++	++		+++
2000 onwards	+++	+++		++	+	+++	+++	+++	+++
Total	13 +	9 +	1 +	5 +	3 +	5 +	7 +	3 +	9 +

Note: For Table 2.1, 9 is considered to be the strongest factor. For Table 2.2, 13 is considered to be the strongest factor.

Key: APFA = Accession à la propriété foncière agricole (APFA: Access to agricultural land) ; PNDA: Programme National de Développement Agricole (NADP: National Agricultural Development Programme).

Source: INRA-SAGA 2010: our fieldwork investigations.

Table 2.3 Land use patterns in El-Guedid district (%)

Land types	*1972 (%)*	*2001 (%)*	*Change (%)*
Sand	21.16	32.88	+11.72
Rangeland	64.63	37.35	-27.28
Cereals	4.85	16.15	+11.30
Uncultivated *dayas* (basins)	9.33	8.43	-0.90
Fallow land	0.00	5.12	+5.12
Urban areas	0.03	0.07	+0.04

Source: SAGA work carried out by the INRA Djelfa 2010: our fieldwork investigations.

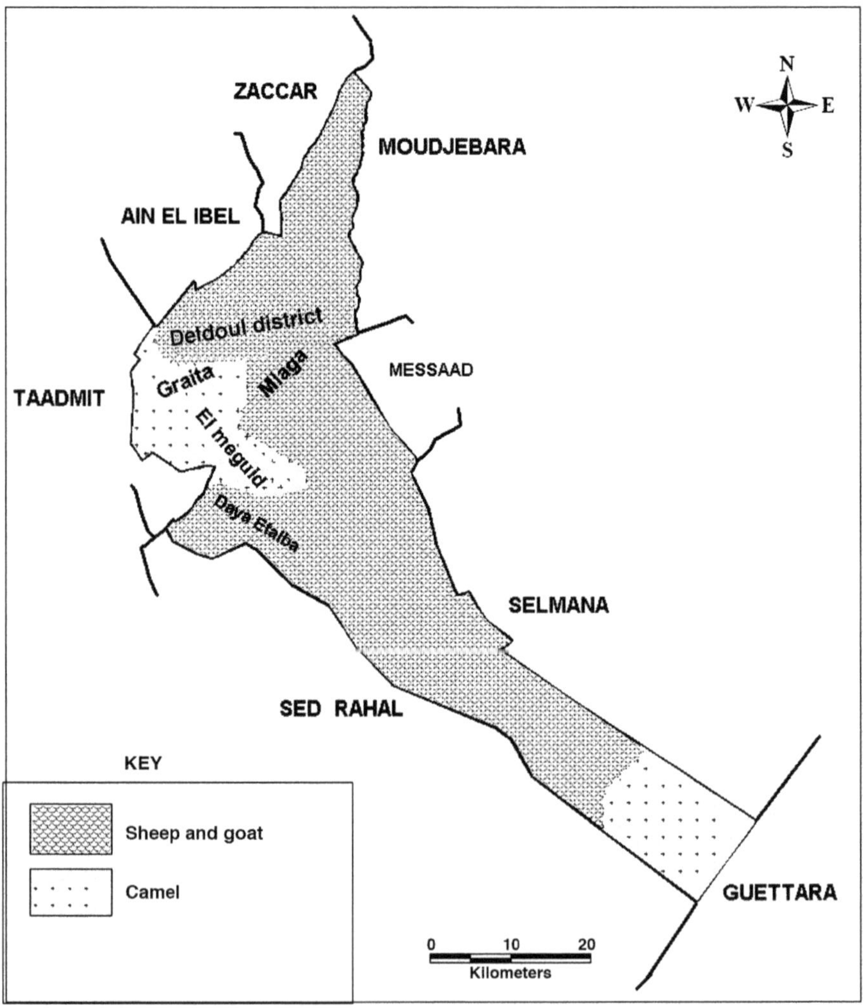

Figure 2.2 Dominant animal-farming systems used in the Deldoul region (Djelfa)
Source: Map created by the group of livestock producers.

Monetised access to feed and the upheaval in livestock production tasks and activities: unequal power of men and women

While pastoral and agro-pastoral populations have managed to deal with the extreme variability of rainfall, this adaptation, often on an individual basis, has however led to the monetisation of pasture fodder resources (natural rangeland, common grazing pasture and cultivated land).[13] The purchase of additional feed for livestock is systematic (grains or feed), even in rainy years. Results show that the practice of this type of livestock production results directly from the fodder shortage on rangelands and a favourable economic situation for livestock producers. Analysis of the livestock market data in the Wilaya of Djelfa reveals that the sale prices for various animals, particularly lambs and yearlings,[14] have increased in recent years (Figure 2.3). Over the last decade, the average price of lambs fluctuated between 80 and 150 euros. This market price allows a livestock producer to purchase at least four to six bags of barley at 25 euros each. This amount of barley can cover the requirements of a ewe with lamb for at least a year in a dry year. It appears that the demographic growth,[15] the increasing demand for red meat and the increased purchasing power of Algerians have tended to increase the ability of livestock producers to deal with the low availability of free pastoral fodder and the new requirements of settling: school fees for children, changing eating habits,[16] etc. Unfortunately, this trend has been to the detriment of pastoral areas as it allows livestock producers to keep their livestock on damaged and overcrowded land much longer, even in bad years.

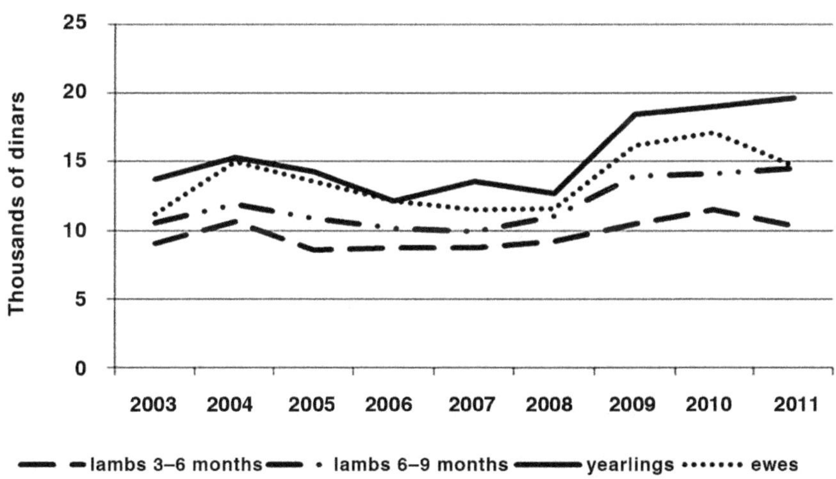

Figure 2.3 Year-to-year variation in prices for small ruminants in the Wilaya of Djelfa

Note : 1 euro = 100 Algerian dinars.

Source: livestock market data INRA, Algeria (2003–2011).

Settling of livestock producers appears to increase women's workload in agro-pastoral farms and the new permanent housing for livestock producers (Tables 2.4a, 2.4b and 2.5). A score of 5 was given to the change in living arrangements. Women are more involved in livestock production activities, especially cleaning, feeding livestock and even herding for short distances around houses. The intensification of livestock production using purchased subsidized feed is likely to increase the demand for women's labor and that may have impacted on time available for other activities. This may reduce the relative economic importance of *qashabiya* given the high value of live animals or meat.

In mobile livestock production systems,[17] women consider the frequent moves a major obstacle to carrying out wool or camel hair processing. The time spent at transhumance sites has decreased significantly and hinders the setting up of the weaving equipment needed to produce the *qashabiya*.[18] A score of 9 (Table 2.1) was attributed to this problem connected to the effects of the decrease in pastoral resources, as this has led to an increase in the frequency of movement. Nevertheless, these related products produced by the women help with the increasingly frequent purchase of animal feed, which is expensive.

Table 2.4a Division of tasks according to production system and gender

	Sedentary		*Semi-transhumant system*		*Transhumant system*	
	Agricultural households		*Agro-pastoral households*		*Pastoral households*	
	W	*M*	*W*	*M*	*W*	*M*
Driving the herds	+++	++	+++	+++	+++	+++
Agricultural work	–	+++		++	–	–
Monitoring and looking after food crops	+++	–	+++	+	–	–
Processing animal-farming by-products	++	–	+++	–	+++	–
Processing (drying and preservation) agricultural products (tomatoes, peppers, apricots, etc.)	+++	–	++	–	–	–
Driving small animals	++	–	+++	–		–
Production of artisan products	++	–	+++	–	+++	–
Housework	+++	–	+++	–	+++	–
Setting up housing (tent, *Kheima*)	–	–	++	–	+++	–
Selling products	–	+++	–	+++	–	+++

Key: + carries out the activity, – does not carry out activity

Source: INRA-SAGA 2010: Deldoul women's focus group (Messaad/Djelfa)

Table 2.4b Timetable of daily tasks for an agro-pastoral household in Deldoul region

Timetable	*Task*	*Person responsible for the task*	*Frequency (D = Daily activity)*
Wake up at 6 am	Preparing coffee	Mother	D
	Housework	Younger daughter	D
From 8.30 am to 12 pm	Weaving	Older daughters + middle daughter + mother	D
	Preparing lunch	Younger daughters/on a rota	D
	Agricultural work	Mother + younger daughters and sons	As needed
From 2 pm to 3 pm	Nap	Only the weavers	D
From 3 pm to 6 pm	Weaving	Mother + middle daughter	D
	Fetching wood	Younger daughter	D
	Preparing dinner	Younger daughters	D
	Agricultural work	Mother + younger daughter + son	As needed
From 9 pm to 11 pm	Television	The whole family	D
From 11 pm to 1 am	Weaving when needed	Middle daughter	As needed

Table 2.5 Constraints scored by women according to their importance (ranging from 1-lowest to 5-highest)

Problem	*Severity*
Weaving expertise no longer being transmitted (ageing of the population)	5
Decline of camel farming	5
Decrease in camel-hair raw material	5
Breaking up of extended family and changing way of life	3
Changing settlement methods	5
Deterioration of natural resources due to socio-economic changes and accentuated by drought	2
Devaluing of product through presence of intermediaries	4
Lack of fees for associations	4
Audio-visual technologies (satellite dish, mobile phones)	4
Low number of souks and selling places	3
Changing animal-farming practices: moving and feeding of livestock	5
Small profit margins	5

Note: a score of 5 is considered to be the most significant factor

On the other hand, in mobile livestock production systems where the herd is entrusted to a professional shepherd, women are completely disconnected from the production sphere, where they were dominant and had a say on the related by-products. This disconnection has made them lose control over the management of livestock resources, especially livestock by-products, and thus related raw material. But this could free them from heavy duties to allow them to work more on *qashabiya*, thus generate more income, even if men are deciding about the use of raw material.

Availability of raw material: local 'wabar',[19] an increasingly rare and expensive material

The information collected from women shows that the greatest problem is the low availability of camel hair to make 'aguiga'[20] *qashabiya*. Sheep wool remains available on livestock markets throughout the year. This is confirmed by our scorings. Scores of 5 were given by women to the decrease in camel hair fiber raw materials and the decline in the number of camels (Table 2.5).

One of the major changes in the livestock production systems is camel farming moving away from the steppes to pre-Saharan rangelands (Figure 2.4). This increased distance from the production centres forces women to buy camel hair at local markets. Indeed, the force field[21] allowed us to identify and better understand the factors behind this situation. Results from the timeline and the livestock production systems map created by the producers allowed us to separate out the events that contributed to the relocation of camels from the steppe regions to pre-Saharan regions. The recent changes to the steppe regions show that the land is being broken up into smaller patches and rangeland is being cut back. Reading from Figure 2.4 and analysing the secondary data clearly shows a strong concentration of camels in Saharan and pre-Saharan regions where there are not many settlements due to the harsh conditions.

Our force field analysis indicates that the uncontrolled slaughtering of young and female animals (makhloul)[22] and the movements of animals to neighbouring countries (Mali, Mauritania, Niger, etc.) are the main factors contributing to the lack of camel hair, especially hair from young camels (Figure 2.5).[23] As a potential solution, participants suggested setting up special brigades responsible of controlling both slaughtering and the borders. In reality these brigades exist but their effectiveness is highly limited.

Access to expertise and related locations: the role and contribution of women's associations

The efforts made by public authorities during the last decade aimed at encouraging a new social category: girls and boys, to perform activities that use the local know-how were not as successful as expected. Indeed, our results from interviews with artisan women and young girls[24] show that the development of training centers has influenced the decisions of young people to look towards other forms of training. The types of training provided were directed to modern occupations, mainly IT,

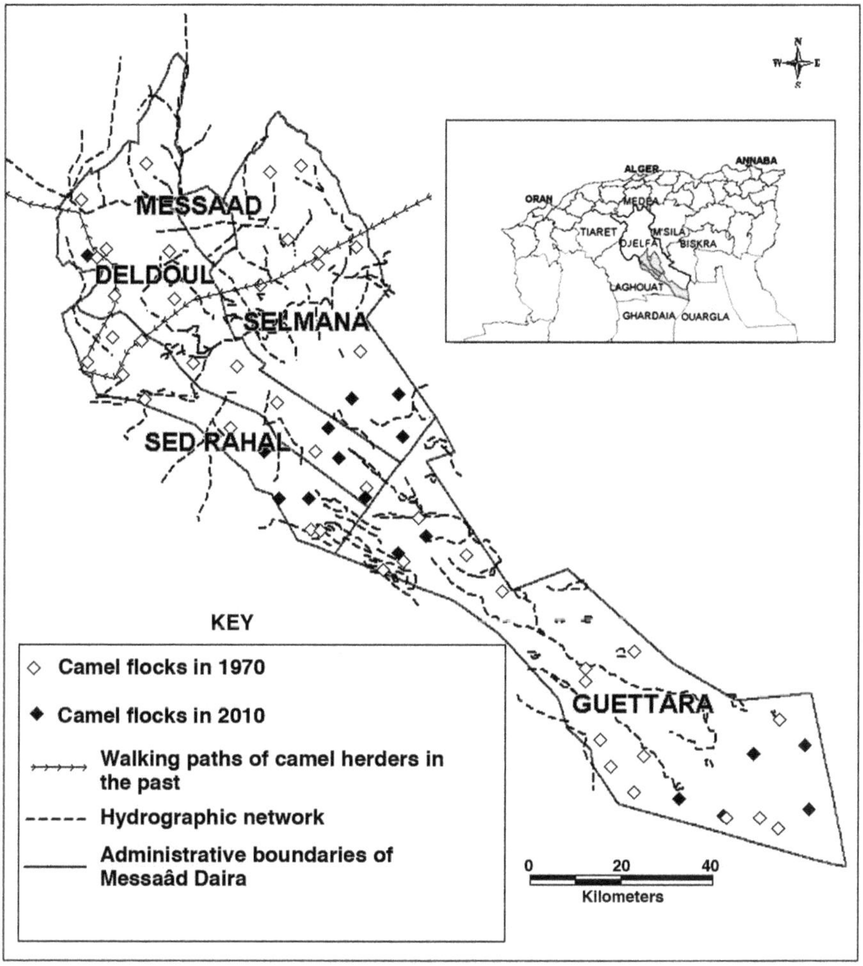

Figure 2.4 Location of camel farming over time (Djelfa)

Note: The following table is part of the map located here

Source: Map created by the livestock producers

hairdressing, building, etc. In reality, decision-makers had not considered the local expertise that could provide income generating activities.

Today, women's associations[25] in rural areas are the preferred learning locations for young girls. This is empowering young girls but the effect is limited due to traditions and customs which prevent women from leaving the home.[26] However, this opportunity has allowed some women to gain a certain amount of liberty and the ability to access information on projects initiated by public authorities. Contact with the outside world has allowed women to participate in economic development schemes and programmes in the same way as men.

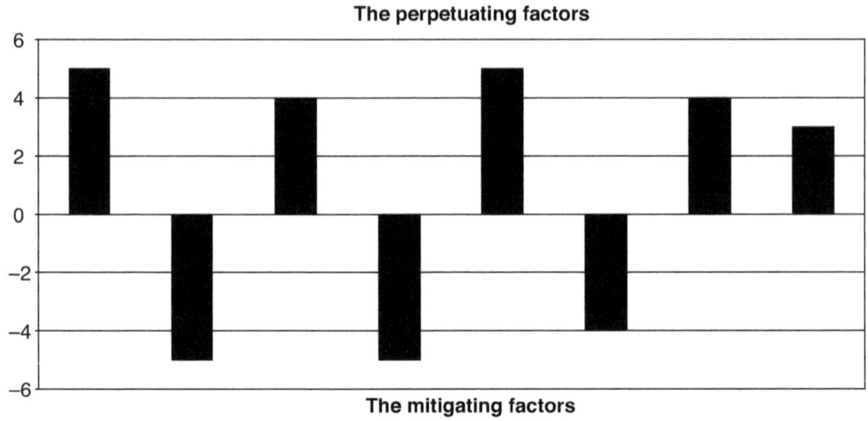

Unchecked slaughtering females	Controlled slaughtering	Escaping across borders	Controlled borders	Movement using vehicles	Benefiting from food subsidies	Difficulty in recruiting shepherds	Drought
5	–5	4	–5	5	–4	4	3

Figure 2.5 Force field analysis: factors behind the reduction in raw materials (camel hair: Wbar)

Women are not entirely out of the decision-making sphere, especially with regard to the means provided by the State to improve rural households' livelihoods. Moreover, it has been noted that this learning model[27] has allowed women to stand out as leaders of women's associations. This is important to consider especially for women and their roles in terms of food security.

Discussion

The consideration of women's and men's views in the analysis has clarified the roles women play in the family economy. In many pastoral regions, the wife is responsible for household food security in the sense that she is responsible for the majority of food processing and storage. Unfortunately, successive droughts, the lack of social services and lack of recognition hinder women's efforts.

A trend towards inequality between women and men: vulnerability of women

Despite the fact that equality between men and women is increasingly being recognised as a key factor in economic prosperity, the deterioration of productive resources has accentuated the inequalities between men and women (Nedjraoui 1981; Grouzis and Albergel 1989; Aidoud 1994; Cornet 2002; Haddouche 2009). Even so, women are involved in even the smallest aspects of daily life and their role in the transmission of local expertise, especially in animal farming and the processing and storage of food, cannot be denied. According to Ouled Salah and Mint Hally (2003), women, through their human potential, are the main producers and play a driving role in family survival strategies and the management of pastoral and natural resources. In Tunisia, women are even involved in collecting pastoral resources to feed animals, especially during droughts (Cialdella 2005). In Morocco, women are much more involved in animal farming activities (Rodary 2007). However, women are rarely involved in making decisions, specifically concerning the use of livestock by-products. Indeed, one of the consequences of this deterioration of pastoral and natural resources is that the decisions taken by the livestock producers, without discussion with their wives, concerning the use of the raw materials for *qashabiya*, prevent women from generating income. Since, in this society of livestock producers in the Algerian steppe, the distribution of income or products of the production sector is often unfavourable to women, that makes them more vulnerable. It appears that despite the efforts made by Algeria since its independence in terms of agricultural and rural development, traditional beliefs have handicapped women in terms of access to the financial resources provided by the public authorities. Moreover, it is only since 2000 that attention paid to women has more clearly translated into development programmes supported by Algeria in rural areas. In reality, the State has deployed tremendous efforts to benefit women in Djelfa from development programmes intended to fund farming activities and projects of craft activities. However, these efforts remain limited. The lack of women's income amplifies existing inequalities between men and women and, as a result, aggravates the vulnerability through the increased scarcity of productive natural resources. This trend is unfortunately an obstacle to poverty reduction in pastoral and agro-pastoral environments.

Consequences of exclusion from decision making: a risk facing animal genetic resources

In old pastoral animal-farming systems, women were actively involved in making decisions regarding the selection of young stock. Indeed, the experience gained through using expertise gave them the ability to select animals with good quality wool and camel hair. Currently, this involvement is falling due to the changes to the animal production system caused by the changes and disruptions mentioned previously, especially the change in land use (Bourbouze 2006). Indeed, studies from projects (SDC-ICARDA-INRA[28] and Mashreq-Maghreb III[29]) reveal the increase in cereals production within rangelands at the expense of natural pastures. Over 10 years, the cultivated land has increased by 136 per cent. Floret *et al.* (1989) and Lhoste (2004) made a similar observation for the south of Tunisia and dry areas of the Sahel: demographic growth and the increased settling of populations resulted in an increase in cultivated land to the detriment of rangeland and livestock. According to Bourbouze and El Aich (2000), these changes in land use are the main factors influencing the decisions by livestock producers to sell their camels and replace them with trucks in Maghreb countries. In our study, decreases in camel numbers have resulted in lack of hair accessibility to women, which in turn has affected the production of specific *qashabiya* types produced from camel hair. As a result of camels moving to desert areas, their hair became scare and more expensive, thus less available than in the past to produce *qashabiya*. Since hair is marketed by men and processed by women, we found that men's work has not been affected because men moved to different geographical areas for marketing purposes as needed, whereas women found themselves with limited access to good quality fiber. They buy camel hair at high prices from local markets, and are rarely involved in decision-making (Rodary 2007; Nafaa, 2002). Moreover, during the interviews a possible correlation was highlighted between animals' adaptation to harsh conditions and their potential for producing good quality wool or hair suitable for processing. The quality of animal by-products (colour and length of wool and hair) is without doubt very important in improving rural people's livelihoods. It affects women more than men due to their vulnerability and their low access to and control over production factors and resources (Joekes and Pointing 1991).

How to reverse the current trends: empowerment of women

Traditions and customs are major obstacles to being able to provide solutions and ensure the financial empowerment of vulnerable women. In pastoral society, crop and livestock production is controlled by men. In our opinion, options that seek to change traditional social and economic relationships to reduce power inequalities between men and women are confronted with almost insurmountable obstacles. Through communication and information technologies, women in rural environments should be equipped and supported in developing income-generating activities for *qashabiya* production. In fact, the Algerian government has set up several funding mechanisms for income-generating projects intended for vulnerable women and women's associations. This type of solution will ensure the transmission and development of artisan products. Moreover, studying the *qashabiya* sector has clearly revealed the increased demand and flows (national and international) for this range of regional products, especially the *qashabiya* from the Messaâd. Indeed, this region is not known for the extent of its pastoral livestock farming but instead for the 'Wabri' *qashabiya* produced mainly by women.

Conclusions

This study has clearly shown that the deterioration of productive natural resources and the consequent changes in livestock production systems have had implications for the traditional women's artisan work of producing and selling products made from wool and camel hair. The changes in pastoral livestock production systems and the in-depth reorganisation of pastoral society have led to phenomena that do not favour the development of the traditional women led income-generating weaving activities.

This research shows that in the transformation of the agro-pastoral society resulting from many confounding factors including the deterioration of productive natural resources, shift to purchased animal feed, and sedentary lifestyle, women have lost economic power compared to the traditional pastoral systems. Indeed, women are increasingly excluded from decision-making, including the use of livestock by-products due to modernisation of processing in factories. As a result, women's income from *qashabiya*-related production activities has dwindled markedly, making them more vulnerable economically and socially.

According to Flintan (2008), the empowerment of rural women can reduce inequalities between men and women in rural areas. However, the specific features and local situation of customs and traditions must be taken into account in order to prevent projects directed to rural women from failing. According to Damamme (2005), traditions, customs and culture are often put forward as reasons for the failure of projects in rural areas.

Notes

1 Institut National de la Recherche Agronomique (INRA), Algiers and Djelfa.

2 International Center for Agricultural Research in the Dry Areas (ICARDA), Amman.

3 Centre International de la Recherche et du Développement (CIRAD), Montpellier.

4 École Nationale Supérieure d'Agronomie (ENSA), Algiers.

5 Institut Technique Moyen d'Agronomie (ITMA), Djelfa.

6 A fraction is defined as a group of families (Boukhobza 1982).

7 The Djamaa, made up of tribe elders, had full control over the management of the rangeland. All tribe members respected this social authority. Indeed, this is one of the reasons that enabled the balance of the ecological system to be maintained.

8 Two rainy seasons are important for the livestock producers in the steppe: October to December and February to April. Outside of these periods, the year is considered to have a precipitation deficit. Indeed, these periods of rainfall are very important for the production of natural and cultivated plants.

9 Generally, rangeland is the property of the state. The cultivation of rangeland and the practice of Gdel (social form of marking territory) are illegal methods of appropriating land, used by tribe members to take over (without a property deed) the occupied land. Today, these practices cause conflicts between tribe members. The land conflicts are more intense during ploughing periods (October to December).

10 Absentee landlords take advantage of belonging to a pastoral community (tribe, fraction or family), using this type of practice and causing considerable damage to pastoral natural resources. Indeed, these people, with no action taken by local authorities, increase the amount of land cultivated in place of natural rangeland and pastures.

11 The rent is very high due to the low pastoral production. The cost of a plot of land of crop residues of 5ha ranges from 500 to 1,000 euros. The cost of the same plot of land rented with green barley ranges from 1,000 to 2,500 euros. Prices are set according to production and the rainfall for the year. Indeed, a year with little rainfall can cause costs to soar as the supply of fodder is low.

12 A major animal farmer is one who has several herds (350 sheep/herd).

13 Property holders that do not have livestock rent their natural rangeland and cultivated areas to the livestock producers. The entire steppe region is shared among the members of the various pastoral and agro-pastoral communities.

14 The yearling sheep, called 'Allouch', are sold mainly during the Muslim ceremonies such as Eid al-Adha religious ceremonies at very high prices. In 2011, prices ranged between 200 and 700 euros.

15 Today, the steppe has around 9 million inhabitants.

16 In the past, livestock producers and their families consumed only livestock products and grains: sheep's milk, traditional butter, etc. Nowadays, they need to purchase staple products: semolina, oil, tinned tomatoes, coffee, tea and clothing for family members.

17 In these systems, the herds are constantly moving. There are two types of system based on the movement of the animals. The first is used by livestock producers without land. They do not have fixed locations or housing. As a result, they are always moving, with all of the family members. On the other hand, the second system differs from the first through the possession of a fixed location where the herd stays after the winter and summer moves. Free and rented pastoral resources are an important factor in their eating habits. The sedentary system is based on very short movements as the herd returns to the place of residence at night. The eating habits are mainly based on the fodder resources purchased.
18 Once set up, the weaving equipment cannot be moved.
19 Local name for the hair from the first fleece of a young camel.
20 Currently the price for fabric made from this raw material ranges from 600 to 800 US dollars.
21 According to Chevalier *et al.* (2008) 'the force field is a component of SAS^2, adapted from a widely used technique in the field of collaborative research'.
22 Camel meat is very popular among the population in the south of the country. Indeed, demographic growth has caused demand for this animal product to soar. Makloul is the local name for the young male and female camels.
23 The price per kilogram for this raw material ranges from 50 to 100 euros.
24 We carried out interviews at professional training centers. This work was centered around the preferences of young female interns in terms of training.
25 According to Kebir (2008), there are 38 women's associations in the Wilaya of Djelfa specializing in traditional weaving.
26 One of the objectives of this study, in order to ensure the continuity of expertise-related activities, is to identify the transmission locations and mechanisms where traditions and customs are not obstacles.
27 Many rural women are already leading women's associations. Within Djelfa, there are over 1,000 women's associations (DPAT 2009).
28 Swiss Agency for Development and Cooperation (SDC) and ICARDA.
29 The Mashreq/Maghreb project began in 1995, and has reached its third phase, entitled 'Developing the Sustainable Livelihoods of Agropastoral Communities of West Asia and North Africa (Mashreq/Maghreb III)'. This third phase is working to improve the livelihoods of poor farmers in low rainfall areas of North Africa and West Asia. In all, the project works with eight countries – Algeria, Iraq, Jordan, Lebanon, Libya, Morocco, Syria and Tunisia.

References

Aidoud, A. (1994) 'Pâturage et désertification des steppes arides d'Algérie, cas des steppes d'alfa', (Stipa tenacissimaL.), *Paralelo* 37(16): 33–42.

Akhilu, Y. (2002) *An Audit of the Livestock Marketing Status in Kenya, Ethiopia and Sudan*, OAU/Inter-African Bureau for Animal Resources, I and II, p. 84.

Bako-Arifari N., and Le Meur, P.-Y. (2001) Une anthropologie sociale des dispositifs de développement, in Baré, J.-F. (ed.) *L'évaluation des politiques de développement, Approches pluridisciplinaires*, L'Harmattan, pp. 121–73.

Boukhobza, M. (1982) *L'agro pastoralisme traditionnel en Algérie, de l'ordre tribal, au désordre colonial*, Alger : OPU, p. 458.

Bourbouze, A. (2006) 'Systèmes d'élevage et production animale dans les steppes du nord de l'Afrique: une relecture de la société pastorale du Maghreb', Sciences et changements planétaires, *Sécheresse* 1(17): 31–9.

Bourbouze, A., El Aich, A. (2000) 'Gestion des parcours et des troupeaux en régions steppiques et réponse à l'aléa climatique', in Guessous, R. and Wageningen, P. (eds)

Livestock production and climatic uncertainty in the Mediterranean, EAAP No. 94, pp. 307–19.

Chevalier, J.-M., Daniel, J., Buckles, J. (2008) *SAS²: Guide sur la recherche collaboratrice et l'engagement social*, : Ottawa : CRDI, p. 364.

Cialdella, N. (2005) 'Stratégies d'élevage dans les projets familiaux en milieu aride – usages des ressources locales pour gérer l'incertain, cas de la Jeffara (sud-est tunisien)', PhD thesis, Institut National Agronomique Paris-Grignon, p. 291.

Cornet, A. (2002) 'La désertification à la croisée de l'environnement et du développement: un problème qui nous concerne', in R. Barbault, A. Cornet, J. Jouzel, G. Mégie, I. Sachs and J. Weber (eds) *Quels enjeux? Quelle contribution des scientifiques?*, Johannesburg Sommet mondial du développement durable 2002. Ministère des Affaires Etrangères/ADPF, July 2002, Paris, pp. 93–130.

Damamme, A. (2005) 'Le genre à l'épreuve du développement au Maroc. Discours et pratiques concernant la place des femmes dans les Projets', PhD thesis, Université d'Orléans, p. 459.

DPAT (2009) *Direction de Planification et de l'Aménagement du Territoire de la Wilaya de Djelfa*. Wilaya of Djelfa monograph, DPAT, p. 300.

DPAT (2010) *Direction de Planification et de l'Aménagement du Territoire de la wilaya de Djelfa*. Wilaya of Djelfa monograph, DPAT, p. 297.

FAO (2009) La parité hommes-femmes dans le secteur de l'agriculture et du développement rural. Guide rapide pour l'intégration de la dimension de genre dans le nouveau cadre stratégique de la FAO, p. 16.

Flintan, F. (2008) 'Etude de la bonne pratique: l'autonomisation des femmes dans les sociétés pastorales. L'initiative Mondiale pour un Pastoralisme Durable', Expert report, UNPD Global Environment Facility, p. 166.

Floret, C., Khatttali, H., Le Floch, E. and Pontanier, R. (1989) 'Le risque de désertification en Tunisie présaharienne. Sa limitation par l'aménagement pastoral', in Eldin, M. and Milleville, P. (eds) *Le risque en agriculture*, Paris : Editions de l'ORSTOM, pp. 291–307.

Grouzis, M. and Albergel, J. (1989) 'Du risque climatique à la contrainte écologique. Incidence de la sécheresse sur les productions végétales et le milieu au Burkina Faso', in Eldin, M. and Milleville, P. (eds) *Le risque en agriculture*, Paris : Editions de l'ORSTOM: pp. 243–54.

Haddouche, I. (2009) 'Cartographie quantitative et gestion des parcours steppiques dans la région de Naama', PhD thesis in Biology (Faculté des Sciences et de la nature, Université de Tlemcen), p. 240.

Hatfield, R. and Davies, J. (2006) *Revue mondiale de l'économie du pastoralisme*, Nairobi : IUCN, p. 54.

INRA (2003–2011) Unpublished statistical database about livestock market data. Algiers, Algeria.

Joekes, S. and Pointing, J. (1991) 'Women in pastoral societies in East and West Africa. Dryland issues', Paper No. 28. London: IIED.

Kanoun, M. (1996) 'Structure des revenus des agropasteurs et effets des techniques de production sur la formation des revenus', Cas de la Wilaya de Djelfa, Masters thesis INA El-Harrach, p. 135.

Kanoun, M., Kanoun, A. and Ziki, B. (2007) 'Les produits d'élevage ovins steppiques: cas de l'agneau de Djelfa', Zaragoza: Seminar of the Scientific-Professional Network on Mediterranean Livestock Farming, pp. 211–17.

Le Houérou, H.N. (1995) Dégradation, régénération et mise en valeur des terres sèches d'Afrique. In L'homme peut-il refaire ce qu'il a défait? de Pontanier, R., M'Hiri, A., Akrimi, N., Aronson, J. and Le Floch H. (eds), Paris: ORSTOM, pp. 65–102.

Lhoste, P. (2004) 'L'analyse des transformations des systèmes d'élevage dans les pays du Sud: questions et perspectives', in E. Chia, B. Dedieu, C.H. Moulin and M. Tichit (eds) *Transformation des pratiques techniques et flexibilité des systèmes d'élevage*, INRA-SAD seminar, TRAPEUR programme, Montpellier, 15–16 March.

Mottet, A. (2005) 'Transformation des systèmes d'élevage depuis 1950 et conséquences pour la dynamique des paysages dans les Pyrénées. Contribution à l'étude du phénomène d'abandon de terres agricoles en montagne à partir de l'exemple de quatre communes des Hautes-Pyrénées', State thesis, INPT-ENSAT, Toulouse, p. 327.

Moulai, A. (2008) 'Suivi de la stratégie méditerranéenne pour le développement durable', Developpement Agricole et Rural Algerian National Study, Vol. 1. PNUE, FAM, PLAN BLEU, p. 44.

Nafaa, R. (2002) 'Femmes rurales Marocaines et développement: inventaire du savoir-faire féminin dans la gestion des ressources naturelles. Faculté des Lettres et Sciences Humaines de Mohammedia, Morocco. Online at: www.beep.ird.fr/collect/bre/index/assoc/HASH017c/f03be2b3.dir/21-375-383.pdf

Nedjraoui, D. (1981) 'Teneurs en éléments biogènes et valeurs énergétiques dans trois principaux faciès de végétation dans les hautes plaines steppique de la Wilaya de Saida', Postgraduate thesis, USTHB, Algiers, p. 156.

Nedjraoui, D. (2003) 'Les mécanismes de suivi de la désertification en Algérie proposition d'un dispositif national de surveillance écologique à long terme', SSO doc., p. 37.

Ouled Saleh, T. and Mint Hally, N. (2003) 'Rôle des femmes Hodh El Gharbi dans l'élevage et la gestion des ressources naturelles', Projet de Gestion Intégrée des Ressources Naturelles de l'Est Mauritanien (GIRNEM), Ph.III. Unpublished project Report 1998–2000.Rodary M. (2007) 'Le travail des femmes dans le Maroc précolonial, entre oppression et résistance. Droit au travail ou accès aux bénéfices ?' *Cahiers d'études africaines* 187–8: 753–780.

Tarhouni, M., Ben Salem, F., Ouled Belgacem, A., Henchi, B. and Neffati, M. (2007) 'Mesure du biovolume de quelques espèces végétales autour de points d'eau en zone aride tunisienne', Sécheresse 18, (4): 240–5.

Vabi, M.-B and Mukong, T.-L. (2002) 'Projet de développement de la foresterie communautaire (PDFC)', Méthode accélérée de recherche participative (MARP). Concepts, principes et applications sur le terrain en gestion communautaire des ressources forestières et founiques: training module, p. 106.

Wilde, V.-L. and Aia Vainio, M. (1995) 'Comment utiliser la méthode accélérée de recherche participative (MARP) pour élaborer les études de cas: Genre et Foresterie', in B. Brindley (ed.) Programme Arbre, Forêt et Communautés Rurales, p. 56.

3 The potential loss of traditional know-how of *qashabiya* production in Djelfa, Algeria

Amel Meguellati-Kanoun[1], Malika Abdelali-Martini[2], Mohamed Kanoun[1], Abdelmadjid Benmebarek[3] and Mohamed Benidir[4]

Introduction

In pastoral societies, women have traditionally contributed to livelihood activities. In Djelfa handicrafts help safeguard an agro-pastoralism in crisis which often does not generate sufficient income. According to 2011 statistics from the Department of small and medium enterprises and artisans, more than 60 per cent of households in Djelfa Governorate currently process wool and camel hair and weave *qashabiya*. This handicraft allows people from marginalised and economically disadvantaged backgrounds to protect themselves against vulnerability. But today, this weaving know-how has become fragile, in part as a result of technical changes due to developments in emerging social, economic and political trends, as well as being affected by overall deep gender inequality which tends to exclude women from any economic participation.

This study is useful in pinpointing the problems that contribute to devaluing this know-how and their causes, and the opportunities for better safeguarding this product. This will serve for better understanding the future of the traditional weaving of *qashabiya* (Figure 3.1) and related gender dynamics and their impact on the livelihoods of populations that depend largely on this product.

The main research question was: what is the current state of the traditional know-how held by women and men and what changes have occurred? Several secondary questions were also posed: a) how do the different existing types of *qashabiya* improve livelihoods? b) what events have contributed to changes in the processes of the know-how, which know-how and how? and c) what potential factors have helped perpetuate the devaluation of local knowledge?

Our hypotheses are: 1) The different types of *qashabiya* help improve the livelihoods of vulnerable households both in urban and rural areas. 2) Despite their major role in the production of *qashabiya*, women weavers are not remunerated fairly. 3) The combination of social, economic and political factors is likely to hamper the continuity of this traditional know-how.

Figure 3.1 Women weaving *qashabiya*: women are the main actors in production of this hand made product

Methodology

Djelfa Governorate is our study area as defined in Chapter 2. We approached Djelfa, Messaâd, Zaccar, Idrissia and Charef, municipalities that are representative of *qashabiya* production, covering both urban and rural areas due to their involvement in the production of this handicraft.

For the sake of analysing gender relations, we organised our work into three stages: 1) consulting secondary data from local institutions, particularly the DSA, DPAT and NGOs[5]; 2) conducting semi-structured interviews (SSI) with the local population, and identifying resource persons within family production units. Questions pertained mainly to the know-how on animal by-products with particular focus on the product in question. 3) Analysis of information and writing up results.

Tools used

We have chosen an approach based on qualitative surveys using a set of SAS[2] (social analysis systems)[6] tools, to identify the various problems, their importance and impact on social changes.

These tools were appropriate to the sociocultural and economic context of the research, including female and male perspectives. These tools were selected only after intensive visits and observations within agro-pastoral family production units using individual and group interviews of livestock farmers, women and men artisans, traders, etc. Individual and group semi-structured interviews (SSI) were the most appropriate in this type of situation.

To retrace the history of the *qashabiya* and understand the changes that have occurred over time and that have contributed to and might have negatively affected the production of this product, the timeline and historical profile tools integrated within the SAS[2], were useful. This allowed us to take stock of the opinions and knowledge of those concerned on the changing situation and the changes that have come about. The historical perspective was supported by recollections and memories of elderly people which were fundamental in creating a full and accurate picture. As such, two groups of people composed of ten or so women artisans aged over 60 were selected to identify events and interpret the facts. These interviews lasted approximately 60 minutes and was coordinated by a trained team of two female researchers and a female technician. The second group was coordinated by a trained team of one male researcher and two male technicians with a group of men composed of eight men artisans varying in age between 50 and 75. To gather information and the changes, we selected skilled artisans/weavers, with the help of resource persons, according to production centres in terms of geographic specialisation. As such, elderly women, housewives and young girls were interviewed[7] individually and in groups by a female researcher and a female technician within the municipalities of Messaâd, Zaccar, Idrissia, Charef and El-Guédid and the results compared and discussed. The interviews lasted between 90 and 120 minutes. We wrote the collected information right after each meeting in the presence of resource persons. This methodological strategy allowed the team to adjust and improve the checklist of questions.

To understand the insights into women's activities in urban and rural settings and understand the weaving and related changes over time, the team held meetings with women living under the same roof from the same household. Two activity diagrams recording the work carried out in terms of *qashabiya* production were drawn up for the Messaâd region according to type of task for two family units composed of women of around the same age (grandmothers, daughters-in-law, girls). In agro-pastoral settings this meeting took place in herding areas around Deldoul, where livestock keeping represents the prime source of revenue, within a larger household. In urban settings the meeting took place in the administrative town with a nuclear family. These highly informative interviews lasted approximately 90 to 120 minutes.

With an aim to understand the problems and direct and indirect effects behind the loss of know-how, the timeline was useful in identifying the problems and their causes by order of importance. Attention was paid to finding the information needed to explain all of the causes and effects. The problem tree revealed how the various players perceive the causes and consequences of the problem in question.

A group of women artisans/weavers from different age groups and various social categories (housewives and public workers, divorcees, widows and single women) and of very different levels of education (illiterate, primary, middle, secondary, and university) from Messaâd aged between 15 and 55 years participated in this exercise.

This exercise lasted nearly 160 minutes and was coordinated by the president of a women's association and a female researcher. The force field analysis was used within the same group of participants in a separate interview, the purpose of which was to identify the factors that are perpetuating the problems and those that may mitigate them. The meeting lasted 120 minutes. Participants were very involved and the tool very useful.

Analysis tools: evaluation matrices

Quantification of the aspects tackled concerning the devaluation of know-how using a scale made it possible to identify and rate the ideas provided by the population according to how strongly they were perceived.[8] We carried out several group interviews (men and women) within the municipalities of Messaâd, Zaccar and Idrissia. The rating and prioritisation matrices combined with the scoring method were easily adopted by the participants, particularly those who were younger. Each exercise lasted on average 30 minutes. The evaluation matrices were used to rate and prioritise the ranges of *qashabiya*, the production centres by type of *qashabiya*, the activities of women in relation to weaving, and the problems contributing to the loss of know-how and the factors that mitigate them.

Results

Historical background to the use of the qashabiya

An item of clothing is a means of cultural expression. The word 'culture' here refers to 'tradition, popular practices and urban life' Grosfilley (2004). The *qashabiya* is a local product for men made using traditional know-how defined by the FAO (2009) as 'a set of historical and cultural practices and traditional knowledge concerning the production and processing of products'.

In order to identify previous and current forms of dressing[9] and be able to plot the changes in this product, the timeline combined with the historical profile revealed three major distinct periods in history: the pre-colonial period, colonial period and independence period (Figure 3.2). Accounts provided by the elderly confirmed that, throughout the pastoral way of life, the burnous (a men's garment of Berber origin) was the prestigious dress of the people during these periods. Its production dates back to the pre-colonial period and the accession of Emir Abdelkader, recognised as one of the key founders of the modern Algerian state and the leader of the resistance against French colonial rule between 1832 and 1847. However, the analysis showed that, in the 1990s, the use of the burnous began to dwindle considerably. This downward trend can be linked with profound

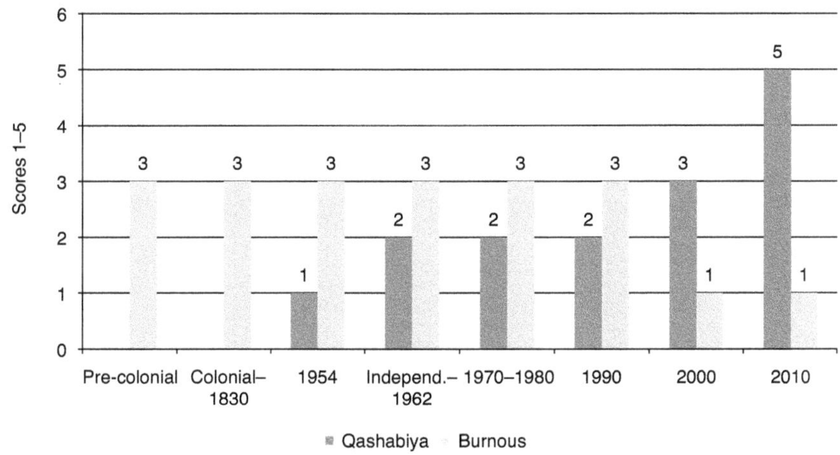

Figure 3.2 Importance of the use of the *qashabiya* and burnous in Algeria (score 1= low, 5=high)

Source: Heterogeneous group interview (men and women)

changes that had affected modes of dressing over the course of socio-economic events, favouring, in turn, a rise in the production of the *qashabiya* aimed essentially at supporting strong domestic and external demand. As such, most of the women artisan weavers interviewed consider this craft a critical one that meets urgent monetary needs, explaining the respective rating of 3 for the importance of wearing the *qashabiya* from 2000 to 2010 (Figure 3.2).

Types and places of origin of the qashabiya*: Djelfa administrative town, a production centre of excellence*

Juxtaposing the techniques of the SAS[2], the timeline and historical profile provided a visual picture of the past and current situations and made it possible to discuss the analysis of the results, and to compare the participants' points of view. A rating scale of 1 to 5 was established and a value of 5 was given by the women artisans having participated in the group interview to the importance of wool, which indicates a significant impact for agro-pastoral populations concerning this material until the 1990s (Figure 3.3). It should be noted that until the end of the 1980s, wool was a highly valued luxury product by the population and mainly used for making domestic consumer goods such as clothes, furnishings and carpets, unlike hair, which was less significant in the pastoral way of life. It was the nomadic production tool, mainly used for making food storage sacks, or grair, and rectangular bands, or flij used for erecting the tent, or kheima, the main means of shelter for migratory livestock farmers. Lower value was given to the economic significance of camel hair until the 1980s (Figure 3.3), a period during which the production

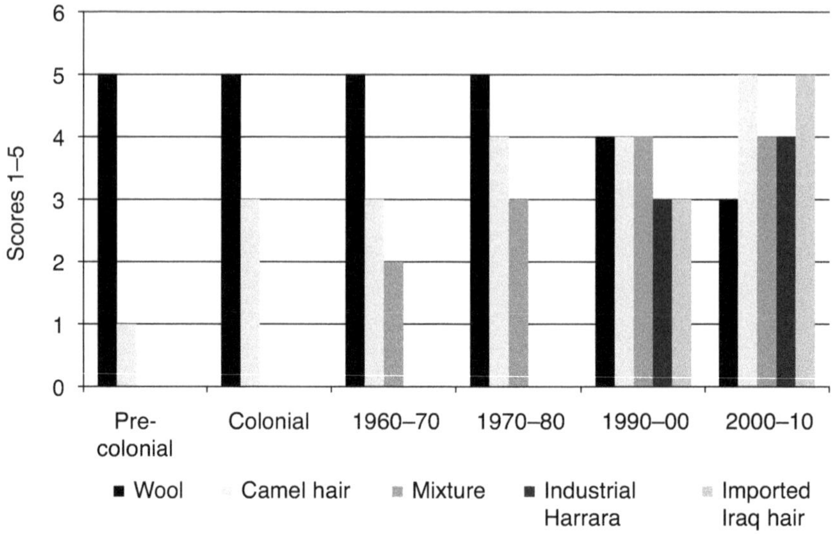

Figure 3.3 Changes in the importance of raw material (wool, camel hair, mixiture, imported 'Iraq' hair, industrial 'harrara') during the pre-colonial, colonial and independence periods

Source: Heterogeneous group interview (men and women)

of the *qashabiya* from hair gave economic importance to this raw material (Figure 3.2) and so the pure camel hair *qashabiya*[10] became a highly 'sought-after' product. The 2000s saw the introduction of new raw materials, particularly pre-washed camel hair imported from the Middle East and the industrial fibre, Harrara. This brings us to consider the importance of the relationship between the availability of the raw material (hair and/or wool) and the know-how of the workers in particular after the country opened up to international trade in the 2000s, consequently promoting the textile sector.

The semi-structured interviews conducted among men and women revealed the existence of a wide variety of *qashabiya* (Table 3.1) under local names attributed either to the colour, nature or mix of fibres of the raw material. These are significant indicators in the eyes of traders and consumers when evaluating the product. A wide variety of local and imported hair and wool products were listed. Three ranges of products were identified under the 'pure' hair, four for the wool and, lastly, two others concerning the industrial fibre (Table 3.1). Differences in terms of specific quality were noted within the varieties of local *qashabiya*.

The combination of tools, such as the social mapping and evaluation and prioritisation matrices (Table 3.2), allowed us to identify the product[11] in question and denote its quality related to its geographical origins.[12] Our analysis reveals that the geographic name boosts the product's reputation. Generally, the populations of Djelfa consider this product as 'theirs' and part of their culture. Specific names,

Table 3.1 Illustration of the ranges of *qashabiya* by raw material

Ranges	*Ouabar camel hair*			*Souf sheep's wool*				*Industrial*	
Type of qashabiya	*Aguiga*	*Adi*	*Lamkhalta*	*Beida*	*Lamkhalta*	*Maqtoussa*	*Draa*	*Fibrame*	*Imported or Iraq*
Local name/ raw material	Nature	Nature	Mix	Colour	Mix	Colour	Colour	Nature	Nature

Source: Homogeneous group interview (women artisans).

Table 3.2 Rating of the production locations of the different ranges of *qashabiya*

Areas	*Hair*				*Wool*						
Ranges of qashabiya	*Camel hair/ Aiguiga*	*Regular/ Adi*	*Imported Iraq*	*Subtotal*	*Lamkhalta*	*White/ Beida*	*Dyed/ Maqtoussa*	*Natural brown/Draa*	*Ind. Fibre Fibrame*	*Subtotal*	*Total*
Messaâd	5	5	4	14	3	1	3	3	1	11	*25
Zaccar	2	3	2	7	1	5	1	2	2	11	18
Djelfa	3	4	4	11	3	2	3	4	4	16	*27
Dar Chioukh	1	1	2	4	1	1	1	3	3	9	13
Idrissia	2	2	3	7	2	2	1	5	4	14	*21
El Guedid	1	1	2	4	1	2	1	3	5	12	16

Note: Scores of 1 to 5 have been given corresponding to the degree of skill necessary in producing the *qashabiya* (1=low; 5=high).

Source: Homogeneous group interview with women artisans.

particularly those used during commercial transactions in the place of origin concerning the white wool *qashabiya* or '*qashabiya* beidha', in other words the '*qashabiya* Zaccar' and '*qashabiya* Messaâd' or '*qashabiya* wabri', confer specific brands with quality and richness in their own rights. The implication is that a product remains closely connected with the local community that created it using techniques and knowledge[13] shared within the local community and used over a long period of time by successive generations in the region.

Over time, and due to the interest in the product, women's knowledge of *qashabiya* production spread throughout the municipality of Djelfa. Divergences spread following migration during the last decade, particularly at regional administrative town level due to a lack of employment and paid work. The group of participants who carried out the exercise concerning the rating of production centres gave a significant score of about 27 to this socio-territorial unit, followed respectively by the areas of Messaâd,[14] and Idrissia with scores of 25 and 21 (Table 3.2). The municipality of Djelfa has become, nowadays, a production centre par excellence with versatile use of a wide range of *qashabiya* products. The reasons are related to a) the high concentration of skilled labour belonging to several pastoral and agro-pastoral communities and b) the proximity to markets (souks) and to suppliers of raw material.

Qashabiya production process

The results obtained through the different individual and group semi-structured interviews conducted with men and women artisans show that women play a central role in the production of the *qashabiya*. Observations concerning the list of tasks reveal that women artisans are largely skilled and are carrying out three quarters of the work (Table 3.3). There are three categories of operations where a number of key tasks are performed either by women or by men. These activities extend from the preparation and processing of raw material to weaving the cloth[15] by women, and finally knitting it, by the men who are the tailors[16] and whose work is likely to be more flexible[17] than that of women.

Production: difference depending on the nature of the raw material

The stage involving preparation and processing of raw material is the most labour-intensive stage for all types of *qashabiya* (Table 3.4). The painstaking cleaning and processing of the raw material seems to be fairly complex for all local camel hair and wool *qashabiya*, requiring physical effort including painstaking dexterity linked with a) particular finger and hand movements, b) skill and observation and c) time and patience.

The results obtained using the evaluation and rating matrices[18] show that the local *qashabiya* called El hora, made from camel calf hair called Aguiga,[19] and white wool, Beidha, are the most difficult to make (Table 3.5).

Table 3.3 *Qashabiya* production process and operations

Operations	*Tasks*	*Sex*
Stage 1		
Cleaning and preparation	Scouring/*Bachm &Tantir* Combing/*Macht* Mixing/*Takhlat 1* Dying/*Quatssa*	Women
Stage 2		
Beaming Warp/processing raw material	Carding/*Kardacha 1* Mixing/*Takhlat 2* Carding/*Kardacha 2* Condensing into strips/*Tkaoussir* Spinning/*Gzil* *Taklasse*	Women
Stage 3		
Weaving 'Nsig'	Winding threads on the warping board/*Sedoua*	Women

Source: Individual interviews and homogeneous group interviews with women artisans

Potential devaluation of traditional know-how and mitigating factors

Discussions with participants indicate that the viability of this handicraft is nowadays compromised due to various economic, social and political problems.

Our careful selection of techniques allowed the identification of the main factors affecting the preservation and continuation of this traditional know-how as well as the trends that may help mitigate its decline.

Demotivating production conditions: unpaid efforts

Poor market access constitutes a major factor hindering the sustainability of this handicraft. Indeed, tradition does not allow all women[20] to take part in economic activities outside the home. Traditional society is still patriarchal[21] and it is down to the man to commercialise the products, except that the sale of the product is generally discussed between husband and wife and/or often the consent of the wife is necessary. As a general rule, women can enjoy the income from their work, which can positively influence their productivity. Moreover, these women state that they do not have any points of reference when it comes to commercial transactions.[22] This is literally called souk ragad which means 'standby market' or 'status quo market': sales prices are always low and do not compensate the efforts undertaken. This absence of regulation in the product's commercialisation gives speculators the opportunity to manipulate prices as they wish. Often the price of a camel hair *qashabiya* is two to fourfold the actual price pocketed by the woman artisan who states that she is forced to 'undersell' in some way her product to meet the household's incurred expenses.[23] This is one of the major reasons for which girls are reluctant to pursue this kind of work and so continue to side line this activity.

Table 3.4 Evaluation of the degree of difficulty of *qashabiya* production per process stage

Ranges	*Hair*			*Wool*				*Industrial*	
Type of qashabiya	*Aguiga*	*Adi*	*Lamkhalta*	*Beida*	*Lamkhalta*	*Maqtoussa*	*Draa*	*Fibrame*	*Imported Iraq*
Processing/ Preparation of the raw material	40	32	32	37	35	22	32	10	26
Beaming warp	17	15	15	17	16	13	16	12	17
Weaving	23	18	18	22	18	13	18	8	22

Source: Group interviews, heterogeneous women (elderly and young) and our calculations

Table 3.5 Evaluation of the difficulty of *qashabiya* production stages

Ranges	*Hair*			*Wool*				*Industrial*	
Type of qashabiya	*Aiguiga*	*Adi*	*Lamkhalta*	*Beida*	*Lamkhalta*	*Maqtoussa*	*Draa*	*Fibrame*	*Imported Iraq*
Evaluation	70	52	52	63	53	43	57	31	54
Rating	1	6	2 6	1 2	5	7	3	8	4

Source: Interviews with elderly women from Messaâd and our calculations

As a result, there is now a significant decline in social solidarity as fewer women get involved in the production process. The processing and production of the fabric requires, depending on the nature of the product, at least two to four women per production unit, which makes the activity a place for social interaction and solidarity. Seemingly, shortage of family labour makes it impossible, as is often the case in practice, to undertake *qashabiya* production, which requires a heavy workload. In pastoral settings, women aged over 70 from the old nomadic system consider the deterioration in pastoral resources, due to worsening climatic conditions, as one of the factors behind the loss of *qashabiya* know-how. The changing socio-economic and bio-physical environments are not favourable to preserving the know-how linked with processing wool and camel hair.

Moreover, according to women, the dislocation of the extended family, considered as the primary economic unit, and the creation of the mono-nuclear family, is another significant factor behind this shortage. The production work still depends on the availability of labour and the amount of work to do on the farm.[24] This is all the more so as the overall significance of the roles attributed to each sex illustrated in Table 3.6 reveals differences[25] by gender in the division of daily and seasonal tasks. This is one of the key reasons why there is a lack of interest in difficult-to-produce *qashabiya* made from camel hair and wool by women at home, who state that they prefer to produce less arduous *qashabiya* from camel hair fibres imported from Iraq and the industrial fibre, fibrame. In addition, another more commercial interpretation was put forward by male participants and young women artisans, where results (Table 3.7) show that their perceptions are more related to demand for the product on the market. Pure camel hair *qashabiya* Aguiga, and products from camel hair imported from Iraq and those made from industrial fibres, fibrame, were given high overall rates of 15, 13 and 11 respectively. Moreover, some divorced women and widows favour the above options due to limited access to inputs, in particular the *qashabiya* fibrame.

In urban areas there is a clear lack of motivation by skilled women labourers to produce local *qashabiya* made from camel hair and wool due to the smallness of the rooms in their homes and their unsuitability for the[26] practice of this activity. This is one of the most important negative aspects indicated by women. It raises questions about the sustainability of this activity in urban areas. But in rural areas this problem can be solved by establishing a facility which meets standards of similar artisan business and meets women's needs. The facility can be developed through grants and managed by a women's cooperative. The cooperative can also open a shop in the village to strengthen its market position.

Tendency toward breaking down the know-how and a shift toward mass production

Nowadays, use of fee-based subcontracting is almost universal, especially in urban areas. The most delicate operations which were practised as part of family tasks requiring know-how in terms of carding, teasing and spinning, and beaming the warp are now remunerated through sub-contracting.[27] This market structuring is quite natural and happens in other industries. But if women are not organized and

Table 3.6 Work conducted by women in the different livestock systems by tasks

Settled and semi-migratory system						*Migratory system*
Type of work	*Agricultural household*		*Agro-pastoral households*		*Pastoral households*	
	W	*M*	*W*	*M*	*W*	*M*
Herding	+++	++	+++	+++	+++	+++
Agricultural work	–	+++		++	–	–
Tending and maintaining food crops	+++	–	+++	+	–	–
Processing livestock by-products	++	–	+++	–	+++	–
Processing (drying and preservation) agricultural produce (tomatoes, peppers, apricots, etc.)	+++	–	++	–	–	–
Small livestock farming	++	–	+++	–		–
Handicraft production	++	–	+++	–	+++	–
Domestic work	+++	–	+++	–	+++	–
Home construction (tent, *Kheima*)	–	–	++	–	+++	–
Sale of products	–	+++	–	+++	–	+++

Key: + practices the activity; – does not practice

Source: Group interviews with women per livestock system

Table 3.7 Selection and preference of *qashabiya* product by gender (sex and age of producers)

Raw material	*Hair*			*Wool*			*Industrial*		
Ranges	*Aguiga*	*Adi*	*Lamkhalta*	*Beida*	*Lamkhalta*	*Maqtoussa*	*Draa*	*Fibrame*	*Imported Iraq*
Men	15	8	7	11	7	6	13	11	15
Women artisans at home	10	13	5	15	9	10	11	11	15
Elderly women artisans	15	14	12	9	12	12	8	6	10
Young female artisans	8	8	5	9	4	5	4	15	11

Source: Heterogeneous group interviews (women and men)

do not improve the practice in the segments of the market they control (facilities, technologies, etc) they will lose even more from this market structuring and they will be open to exploitation as traders try vertical integration and women lose control of some parts of the production. Married women, widows and divorcees interviewed, whether exercising a professional activity or otherwise, consider this activity as a means of freedom and time saving, helping them to use their labour more effectively for household work and for their children's education. In the face of this new trend, elderly women revealed a certain 'discontentment', asserting that this solution and change in the social division of work through the introduction of a group of subcontractors could jeopardise the future of the *qashabiya* due to the medium-term break in the continued transmission of know-how. It is true that the lack of control over all of the operation by the producing women artisans may even contribute to and affect the deterioration in the product's quality. That is why women insist that all the production work must be carried out within the family unit to guarantee quality and the apprenticeship of young girls[28] starting at the age of seven (Table 3.8). Nevertheless, the policy of socialisation and access (for an increasing portion of the female population[29]) to education and training has, in some way, impacted on this handicraft work given that the production programme, which generally depends on commercial criteria,[30] is subject to tight time constraints and the time of year.

Table 3.8 Methods of transmitting the various *qashabiya* production tasks to young girls by degree of difficulty

Operations	*Cleaning* Bachm	*Mixing* Takhlat	*Carding* *Tkardiche*	*Spinning* *Gzil*	*Beaming* *warp Sedoua*	*Weaving* *Nsig*
Degree of difficulty	1	2	3	5	6	4

Key: Order of difficulty from 1 to 6 (6 = most difficult)

Source: Homogeneous group interviews with elderly women artisans

Results from the force field analysis clearly show that the factors perpetuating the fear of loss are powerful and reflect an extreme vulnerability due to the ageing of the population who holds this know-how, the loss of interest by young girls and the lack of motivation of women artisans faced with overwork and physical exhaustion that is unfairly and unequally paid. According to the participants' accounts, this situation may worsen to the point of depriving them of an income that meets their subsistence requirements in an environment where employment opportunities are extremely low. Direct links between factors that perpetuate the problem and those that mitigate it were established and prioritisation from 1 to 6 given below for:

1 Providing financial support and making micro-credit interest-free and accessible for purchasing raw material (5).[31]
2 Developing technical skills of women at home (4).
3 Providing unmarried women artisans (young artisans and widows and divorcees) with shops in commercial centres (4).

4 Reducing production workload and costs of subcontracting with technological support: providing women with electronic spinning equipment, maghzel, and weaving equipment appropriate for flats (4).
5 Linking training with qualifications within associations and training centres to get young people interested (3).
6 Facilitating the delivery of business cards (3).

Discussion

The analysis revealed elements concerning the main factors that threaten to damage this heritage along with some potential solutions. In pastoral areas, women are the main contributors[32] to household food security in the sense that they are responsible for several 'on farm' activities. However, successive droughts, insufficient social services and lack of recognition are elements that hinder the development of handicrafts and the empowerment of women with regard to income-generating activities.

Worrying results: potential devaluation of traditional know-how

The analysis clearly reveals a loss and devaluation in the production of *qashabiya*. Its product rating according to its producers varies depending on raw material (hair, wool, industrial fibres) and the degree of arduousness required versus the low marketing added value, in addition to the risk of discontinuity in the transmission of know-how due to a lack of succession and disinterest by young people,[33] who nowadays are more inclined towards services and competitor professions.

This is compounded by the fact that, in such traditional societies, the man, whether father, husband or brother, is always at the top of the family hierarchy holding unlimited power[34] and he is in charge of marketing all products. This illustrates prioritisation in the household tasks.[35] Rodary (2007) rightly notes that women's subordination and dependence to their male relatives is specific to the whole of the Maghreb. Women artisans/weavers cannot participate in economic activities outside the home. That is one of the reasons behind the low prices by which their products are sold, sometimes hardly covering production costs. This is due to the lack of information between producers and buyers. In this respect, the analysis clearly shows that the lack of market access hinders women from playing a stable and continuous role in this activity. The *qashabiya* sector remains complex and poorly defined, it is experiencing many problems in terms of control and the determination of a fair income. To conclude, it would indeed seem that women producers are confined to a very low added-value activity. Dammane (2005) notes in a study conducted in Morocco that women's products can only find an economic outlet through a low sale price. This differential implies repercussions and consequences in terms of commercial reliability and profit made on the products to the benefit of intermediaries and traders.

The problem of migration to cities is continuing due to unsatisfactory living conditions (unemployment and poverty) which may lead to a loss of labour. But this does not reduce the accelerated rural exodus and migration of entire pastoral families to cities, further leading to rapid and chaotic large-scale urbanisation as also reported by Ben Ali (1991 in Rodary 2007). This has in turn accelerated the process of social change within village communities towards a situation of increasing emigration by men[36] and of increasing workloads for women with sedentary lifestyles.

With persistent droughts, the migratory flows of men and underemployment in rural areas, the attitudes toward women's work have started a positive change. As a consequence, women's emancipation is at the heart of the debate and the role of development policies must place emphasis on promoting women's self-employment in handicrafts, to generate better income and protect the indigenous know-how. The collected information provides a basis for a strategic and action plan to achieve the objectives of the value chain. The stability of this activity could be improved by starting to develop gender awareness and ensuring an equitable participation of all the chain players in decision-making. This could be done through the involvement of different professionals from other sectors (public bodies, tourism, schools, etc.) within regional cooperatives and associations. Emphasis must also be placed on training by creating professional training centres in various areas per specialised geographic centre in order to attract people's interest in this sector. Equipped training rooms for producing *qashabiya* should be made available to young people in schools, secondary schools and upper secondary schools. This could be a good start to giving young people some basic information on income-generating activities from livestock by-products. It may even offer an alternative security for university graduates who are long-term unemployed.

Advantages and disadvantages of participatory approaches applied

By offering everyone the opportunity to actively participate and express their own views, the participative tools allowed us, as mentioned by Chevalier *et al.* (2008), to identify key events likely to explain this situation. Overall, it appears that certain stages should receive particular attention. It would be essential to carry out an analysis of the environmental conditions and assess the population needs, which would trigger and facilitate gradual awareness about the environmental status versus the issue tackled in this study. It would constitute an absolute prerequisite for the success of subsequent stages. However, the active participation of the population and the use of participative tools cannot necessarily substitute formal surveys, although using these approaches made it possible to answer the research questions and assumptions aimed at understanding the attitudes, perceptions and opinions of the members of the community. The work is being completed by a formal survey conducted among rural households to quantify certain pertinent aspects of our analysis.

Conclusion

This research carried out among local communities in Djelfa revealed the importance of the traditional *qashabiya*, a handicraft made from livestock fiber such as wool and hair that is nowadays in great demand. A combination of socio-economic and biophysical factors hampers the preservation of this handicraft as a result of exclusion of women from market access despite their heavy contribution to household livelihoods through its production.

The *qashabiya* value chain is undergoing a transformation that could negatively impact on the portion that rural women control in this market. The most difficult and labour intensive operations are cleaning, carding, spinning and beaming the warp, they require time and space, and have become monetised without guaranteeing the quality of the *qashabiya* produced.

The harmful effects of industry include availability of low grade raw material, namely industrial fibre and imported pre-washed camel hair. These are not fit for producing the valued traditional quality *qashabiya* and thus devalue both the product and the know-how used to produce it. The transformation of the industry seeking cheap material and more efficient production processes may exclude the local *qashabiya* products with specific qualities made from pure hair, the Aguiga, and white wool, the Beidha, with implications for local knowledge and livelihoods of many rural households relying on that sector.

Analysis shows that women should be well-equipped and well-supported to develop income-generating activities from handicrafts such as the *qashabiya*. Gender-biased traditions and customs constitute a considerable obstacle that must be overcome to provide opportunities for financial autonomy of women.

Notes

1 Institut National de la Recherche Agronomique (INRA), Djelfa.
2 International Center for Agricultural Research in the Dry Areas (ICARDA), Amman.
3 École Nationale Supérieure d'Agronomie (ENSA), Algiers.
4 Unité de recherche Sétif INRA (Research Unit), Algeria.
5 DSA: Direction des services agricoles (Department of agricultural services), DPAT: direction de planification et de l'aménagement du territoire (Department of planning), NGO: non-governmental organisation.
6 These are participative research methods and continuous learning processes for a better understanding of the conditions and problems of the local populations. As the two approaches place emphasis on multi-disciplinarity and on interaction between the researchers and the local population, this interactive and multidisciplinary process is a source of enrichment in that it allows the active sharing of experiences and points of view. However, the two approaches demand know-how in communications and opinion sharing (Wilde and Aia Vainio 1995; Vabi and Mukong 2002; Chevalier *et al.* 2008).
7 Perfect command of the local dialect is a significant asset in this type of research.
8 This is an odd-numbered rating scale from 1 to 5, with 5 being the highest, 3 and 4 average and 1 and 2 the lowest.
9 The djellaba or *qashabiya* and the burnous made from camel hair are traditional local products of the Ouled Nail (APS 2011).
10 The idea of making the *qashabiya* from pure hair came after the production of the first burnous from pure hair for the former President of the Republic, Houari Boumediene.
11 A local product is based on a system of interactions between a physical and biological setting where a set of human factors is involved to express an originality, confer a specific type and thus make its reputation (FAO 2009).

12 These typical production features reflect a legacy which, according to the FAO (2009), has historical and geographical origins rooted in a region through a heritage and a cultural identity.

13 In this case, specific to the processing of the raw material, developed over the years based on local genetic resources, particularly breeds of animal and quality of the material.

14 In the city of Messaâd, 83 per cent of families support themselves through this type of handicraft (APS 2011).

15 Workload calculated according to criteria given by the population on the basis of the total physical effort, experience and the time required by type of product. These criteria depend strongly on the type of *qashabiya* and the type of production stage in question.

16 It must be noted that, nowadays, this category of artisans from a higher age bracket and with low financial capital is almost entirely small scale and continues to practice in small traditional shops with scant and old equipment.

17 Assembly of the product depends on the type of *qashabiya* ordered and the time required is between 15 and 30 days depending on whether it is a hand-sewn or machine-sewn *qashabiya*.

18 It was the women who evaluated the degree of arduousness according to their personal perceptions. Our task merely involved evaluating the rating given by the women. Arduousness is shown with high scores.

19 *Qashabiya* produced from the hair of young camels is similar to down (the animals' first covering).

20 Some women from a certain social category, widows and or divorcees over the age of 60 are allowed to participate.

21 The man, whether the father, husband or brother, is always at the head of the family hierarchy.

22 Despite this obstacle, the woman artisan-cum-trader still tries to understand the interests of the market before, during and after production with the aim of seizing suitable opportunities and avoiding 'dishonest' speculation by traders concerning her product.

23 Debts are incurred particularly to buy raw material.

24 As part of the survival strategies developed by the government in rural areas, it would seem that women are much more involved than previously and brought to contribute in more varied ways.

25 Women confirm that for them 'time is a rare commodity' due to the exercise of several concomitant activities, namely self-subsistence family farming and livestock farming, notably sheep, cows, goats and poultry.

26 The established architectural model features a lack of rooms (1 to 2-bed flats), of small size (3m by 2.5m) and with poor ventilation, which poses problems for the exercise of this activity, such as a nuisance to the neighbours, the location of the room, the cleaning of the raw material, which may result in the spread of mites and cause allergies and respiratory illnesses.

27 With the decline in available labour, opportunity costs are becoming more widespread, particularly for pure hair in terms of the cleaning (bachm) and processing of the raw material (carding and garnetting) in addition to beaming the warp (sedoua). Prices are in the region of 12,000 Algerian Dinar for pure hair, 4,200 for Beida, and 4,000 for Iraq hair.

28 In Table 3.4, the operations that must be passed on to future women artisans are illustrated and rated in order of priority according to their level of difficulty. Acknowledging the value of the processes can lead to the safeguarding of the product.

29 Learning and transmission is more problematic for girls in cities who have better access to schooling.

30 Which implies that the woman artisan-cum-trader must work fast and make her product quickly, especially when winter or a festival period is approaching.

31 Each mitigating factor was given a score from 1 to 5 by the participants, with five being a high score.
32 Nevertheless, the data recorded by the ONS (office national des statistiques; national statistics office) cited by Medouni *et al.* (2006) show that a low contribution of just 1% of women work in farming professions and confirms that just 3% of women exercise an economic activity in a rural setting. Statisticians, economists and even experts count them as unproductive (CNEAP 1999), however, and still according to Medouni *et al.* (2006), this data does not reflect reality given that we are in a world where economic value is calculated in monetary terms.
33 Nowadays, the exercise of such an activity reflects social reluctance on the part of young people, despite recognition by the public authorities, aiming to secure a 'stable' job in economic and service businesses. This may in fact explain their free circulation in an urban structure different to the institutional framework.
34 Even the decision by women to exercise such an activity requires prior agreement by the husband.
35 Doris and Zimprich (1998) point out that the roles and working conditions are not always static but are constantly being redefined according to social changes.
36 This is a structural development that does not totally challenge the traditional power or influence of men in the decision making. However, these changes are accompanied by a positive development in attitudes concerning the perception of the economic role of women and the attitude of men vis à vis productive activities.

References

Chevalier, J.-M., Daniel, J. and Buckles, J. (2008) *SAS²: Guide sur la recherche collaboratrice et l'engagement social.* CRDI, Ottawa, p. 364.

CNEAP (1999) Centre national d'étude et d'analyse pour la planification. Femmes en activités. Lettre Numéro 10, p. 4.

Dammane, A. (2005) Le genre à l'épreuve du développement au Maroc/géographie, Aménagement, Environnement Discours et pratiques concernant la place des femmes dans les projets. Thèse de doctorat, université d'Orléans, Orléans.

Directorate of Agriculture, Ministry of Agriculture and Agrarian Reforms (2011) Unpublished database.

Direction des services agricoles (DPAT) (2008) *Statistiques agricoles.* Cahiers statistiques, Djelfa, p. 180.

Doris, G. and Zimprich, E. (1998) *Prise en compte de la dimension du genre dans le secteur post-récolte Conseils utiles pour l'identification d'aspects spécifiques au genre dans le secteur post-récolte.* Deutsche Gesellschaft fuer Technische, Zusammenarbeit (GTZ) GmbH.

Eudoxie, S. (2006) *Thème: Genre et pastoralisme.* Bulletin trimestriel d'information genre du Réseau Billital Maroobé, Ouagadougou.

FAO (2009) Parités hommes femmes dans le secteur de l'agriculture et du développement rural. FAO, Rome.

Filali, M. R. (1994) *Femmes et travail.* Le Fennec, Casablanca.

Grosfilley, A. (2004) Citation extraite de l'ouvrage L'Afrique des textiles. Edisud, Aix-en-Provence.

Kergoat, D. (2000) Division sexuelle du travail et rapports sociaux de sexe, in Medouni, Y., Boulahchiche, N. and Brahimi, R. (2006) Rôle de la femme rural dans les systèmes de productions agropastorales. Cas de la fraction de *Ouled Baida* de la zone d'El-Guédid. Région de Djelfa (steppe centrale), option Méditerranéens, série, Numéro. 70.

Ministère des finances du Sénégal (1997) Enquêtes sénégalaises auprès des ménages ESAM Rapport de synthèse, Dakar.

Nussbaum, M. C. (2000) *Women and Human Development: The Capabilities Approach.* Cambridge University Press, Cambridge.
Ouoba, R., Tani, M. and Toure, Z. (2003) Analyse stratégiques des enjeux liés au genre au Burkina Fasso. Rapport, Ouagadougou.
Rodary, M. (2007) Le travail des femmes dans le Maroc précolonial, entre oppression et résistance. Droit au travail ou accès aux bénéfices. *Cahiers d'études africaines,* numero 47, Décembre.
Smail (1999) *Aspects de l'aménagement de la steppe algérienne cas de la wilaya de Djelfa.* Thèse de doctorat, université Paul Valery Montpellier III, Montpellier.
Terrillon, J. and Sangaré, N. (2009) Analyse de genre du secteur des organisations de la société civile au Burkina Faso. Programme de renforcement des capacités des organisations de la société civile (pros), Rapport, Ministère de l'économie burkinabé, Ouagadougou.
Vabi, M.-B. and Mukong, T.-L. (2002) Projet de développement de la foresterie communautaire (PDFC). Méthode Accélérée de Recherche Participative (MARP). Concepts, principes et applications sur le terrain en gestion communautaire des ressources forestières et fauniques, *Module de Formation,* p. 106.
Wilde, V.-L. and Aia Vainio, M. (1995) Comment utiliser la méthode accélérée de recherche participative (MARP) pour élaborer les études de cas, *Genre et Foresterie. Programme Arbre, Forêt et Communautés Rurales.* Editeur, Bonita Brindley, p. 56.

4 Gender and marketing of *qashabiya* in Djelfa (Algeria)

Abdelmadjid Benmebarek[1]*, Mohamed Larbi Cherfaoui*[2]*, H. Ouzane*[2]*, Malika Abdelali-Martini*[3]*, Amel Meguellati-Kanoun*[4]*, Mohamed Kanoun*[4] *and Souhila Fodil*[5]

Introduction

To ensure sustainable livelihoods and reduce rural poverty, rural development is the obvious path to pursue. Given the rapidly growing population, it is important to increase access to food, natural resources and various services. In a world of limited resources, the starting point should be improved management of natural resources. Rural development also means paying the necessary attention to territorial governance while at the same time establishing effective links between policies, programs, institutions and populations.

The question addressed here is the organization and function of the marketing channels for local products, especially the by-products of livestock farming. We specifically focus on *qashabiya*, a local garment made of wool and camel hair. In Algeria very little information is available on the marketing value chain for this product, the respective roles of men and women, their marketing practices or the distribution of the added value generated. This study attempts to fill some of these gaps. Historically, local products remained marginalized during the post-independence period and up to the beginning of the twenty-first century, but are now on the policy agenda, particularly under what is called 'rural renewal'. Also, artisanal activities provide a significant additional source of income for women and men in rural households, especially smallholders and agro-pastoralists. Furthermore, social norms and values which determine the specificity of social relations dictate the social dynamics in rural areas with a clear dominance of men to the detriment of women.

To date, sectorial policy of rural development followed a 'macro' approach. Therefore, this study aims at filling in the gaps to bring the necessary improvements to the sectorial policy and facilitate the setup of more pertinent programs and projects.

At the regional level of the Maghreb, local products are embedded in the social life and their survival is hugely importantly to the community, despite the shocks resulting from the impact of modernization and technology (the opening up of national markets, cultural, economic and social shocks induced by alternative products perceived as modern, etc.). These countries have initiated policies aiming at protecting and capitalising on artisan knowledge and know-how, at different periods and with different impacts.

In terms of traditional dress at national and local levels, Algeria includes major zones where the *qashabiya* is worn. It is dominant in the southern slopes of the Saharan Atlas, a dominance that increases as you move from east to west across the country. This localization corresponds on the one hand to the areas of sheep, goat and camel farming and, on the other, to the sociocultural history of each of these regions (Ageron 1971; Boukhobza 1982; Benachenhou 1980). *Qashabiya* of high quality is also worn as ceremonial dress.

In recent decades Algeria has experienced major changes in its value systems that have resulted in the rehabilitation of a number of local products, in particular the *qashabiya*. Whereas it was previously considered at the bottom level of the social hierarchy, the *qashabiya* is today positioned as a prestigious product and is a symbol of the status of its owner through its association with a particular social group.

The factors at the origin of this development appear to be linked essentially to the resurgence of the values of the rural world, strengthened by the rise in religious belief, especially during the winter and periods of religious worship (Ramadan for example). This situation is giving rise to increased demand for this product, which is stimulating its production and the emergence of new informal marketing routes, especially outside mosques.

Research questions

The literature on the production and processing of local products discuss mainly employment and income aspects of rural households. However, the contribution of women in the value chains for these products and the constraints they encounter to access markets is also an important aspect. But the tradition creates a virtual monopoly of men in the market for the product. These traditions and customs remain very strong, and they impact on the value chain as a whole but more particularly on rural woman.

Our research questions are: 1) who are the principal actors disaggregated by gender involved in the value chain of *qashabiya* – from production to marketing? 2) What are the gendered power relationships within the value chain and their impact of the final product?

We posed two hypotheses in the research. The first is that women, in spite of being the principal actors in *qashabiya* production, play only a very marginal role in marketing the finished products due to their limited market access. The second hypothesis is that the unbalanced power relationships between the actors of the value chain undervalue the work involved in the production of the product. The inequitable sharing of the added value generated to the detriment of producers and artisans constitutes a significant constraint facing the preservation of knowledge and know-how. This is further aggravated by the effects of the free economy and globalization.

Methodology

We selected the region of Messâad (Wilaya[6] de Djelfa, Algeria) as the study area. It is described in Chapter 2. Our choice is driven partly by the high potential of this agro-pastoral area for the development of local products, and also because Messâad is recognised as the bastion of production of the high-quality *qashabiya* in Algeria.

We have used semi-structured interviews (SSIs), force field analysis, classification matrices and a timeline to identify and understand the different actors of the value chain. Power relationships were understood through the use of SSIs taking into account different production chain cycles, in addition to the social analysis CLIP (collaborative, conflicts, legitimacy, interests, power). To assess the constraints facing women producers, we used the SSIs at markets, with other key informants, and mapped the results into diagrams. Finally we used the SWOT tool to identify the opportunities and entry points for potential solutions.

Results and discussion

The main focus of this analysis is to unveil the strategies used by women and men for marketing *qashabiya* and compare our results with others from the literature dealing with similar products. We start with a description of the structure and general organization of the *qashabiya* value chain; starting upstream. We analyse the different aspects of marketing and related practices and their impacts knowing that these have experienced major changes in the last four decades.

Figure 4.1 shows that upstream of the value chain, the first are agro-pastoralists, who have wool and goat hair obtained in the region, and camel hair obtained principally from the wilayas lying further south. Camel rearing is declining, due in particular to climate change and its effects on the rangelands, which have clearly decreased in certain areas, and also to insufficient interest and support from government prior to the policy of 'Renewal'. Parts of the wool and camel hair are either used on farms for weaving, or are given away or informally bartered between friends and relatives. The rest are sold by agro-pastoralists or by the shepherd to traders, who themselves sell the product to other wholesalers. A second marketing route is that of the collectors who come from other wilayas.

The *qashabiya* market and related practices are still driven by customs. It can be noted that the production process involves females working at home, males working at their workshops, as well as women recruited to carry out various wool and camel hair processing operations, for which they are paid on a piecework basis (Chapter 3). Marketing takes place after the *qashabiya* is produced; it can take various forms.

On the relative weight of commercial practices, *dallal*[7] is the first contacted to sell *qashabiya*, followed by the wholesaler (Figure 4.2). The retailer is on a par with the family member, sales from home come as a last resort. The numbers of Dellals in Messaâd market runs into hundreds and are all men, whereas large wholesalers are less than a dozen throughout the region and do business with major operators in other regions as well as with foreign operators.

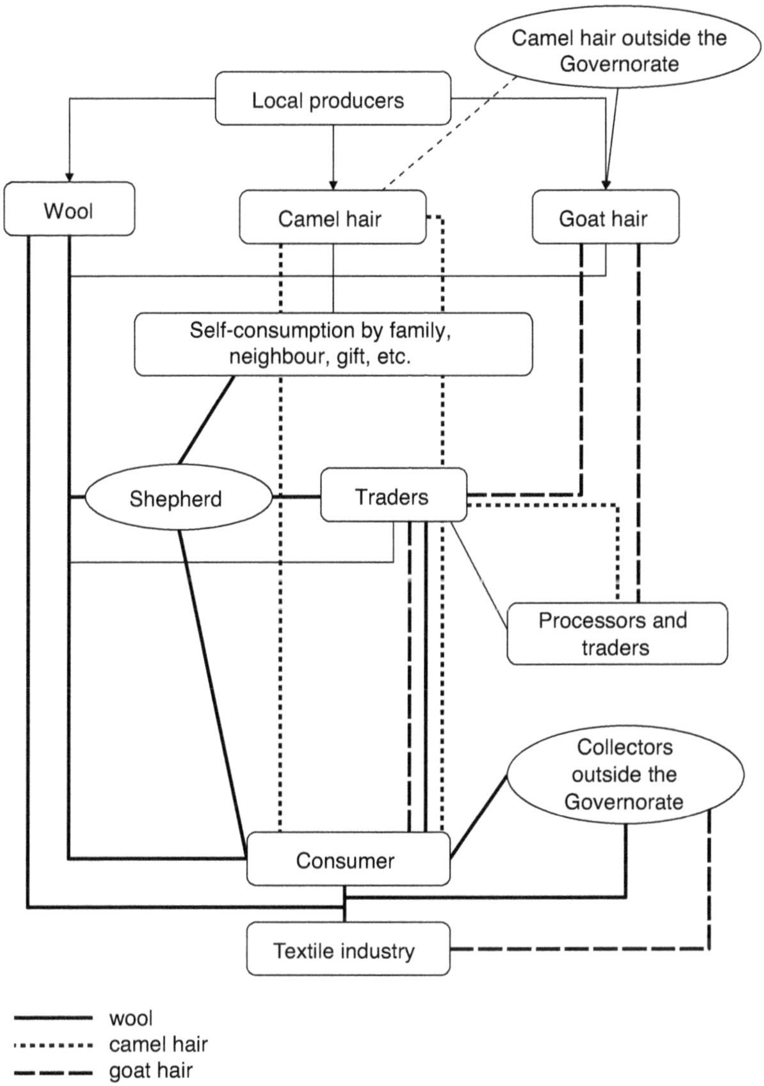

Figure 4.1 The actors and routes taken by the raw materials needed to produce the *qashabiya* products

A previous analysis of marketing roles indicates that, paradoxically, the 'chaabi' – a 'man', who does not show any particular characteristic apart from originating from the region and being a member of one of its communities – stands in first place, with a weight above that of the local trader (2nd position), followed by elderly women and girls aged under 15 (Figure 4.2). This clearly shows that women artisans are excluded from any direct involvement in the commercial sphere. This lack

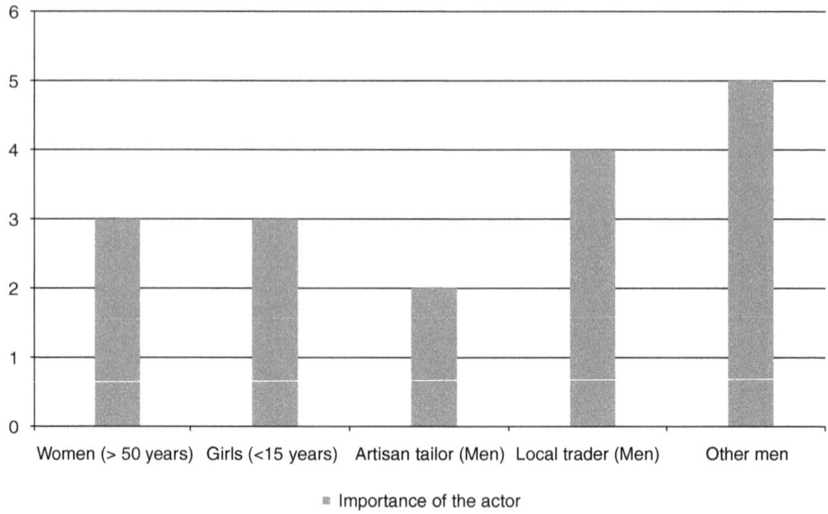

Figure 4.2 Classification matrix per relative weight of commercial practices by actors

of market access is most probably mainly due to the still very strong traditional social norms that assign a strictly determined role to women, usually confined to the home. Women are thus unable to ensure a full valuation for their products on the marketplace. In addition, this lack of market access is also a constraint facing various government and professional institutions and structures, in particular Agriculture and Rural Development, Artisan, Interior and Local Authorities, Solidarity, Commerce, Chambers of Agriculture and Crafts, APC (municipal council), banks, Taxation, and the related operational structures). As a result, women producers appear impeded in their movements and initiatives and are unable to fully assume their roles as economic actors who generate wealth and contribute to the improvement of livelihoods. The only women and girls who access markets are those who act as 'representatives' of women artisans. However, we are now witnessing a considerable improvement in the access of female artisans to information on market prices and to more general developments about marketing, through the dissemination and rapid adoption of modern forms of communication (mobile phone) that allow female producers to be in direct contact with the commercial operator.

A perfectly organised marketing system that is integrated at local, domestic (Wilaya of Djelfa) and international (Maghreb, Arab world) level

Marketing channels of qashabiya

We examined the marketing channels (Figure 4.3) from a spatial point of view, while also taking into account the categories of actors concerned by the product flows – whether in unfinished or finished form ready to be worn. Analysis of the

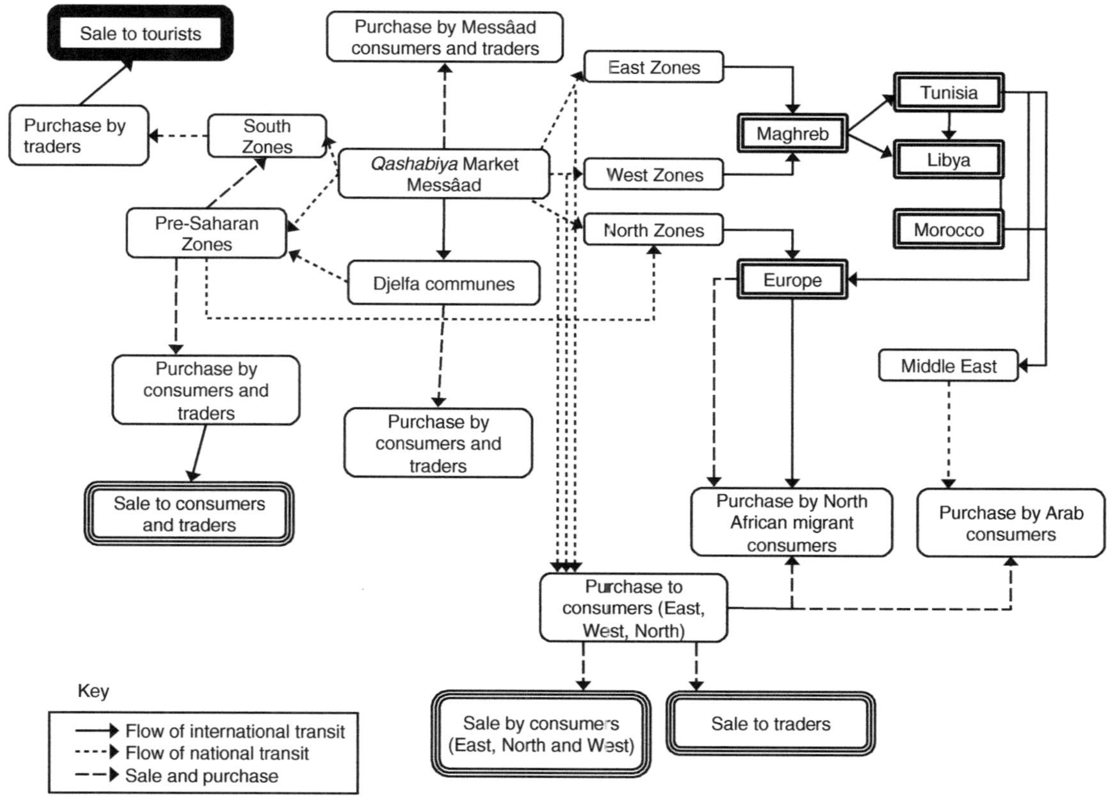

Figure 4.3 Marketing channels and types of actors involved in the Messaâd *qashabiya* flows

information collected yielded a number of conclusions. First, there is the linchpin role played by the Messaâd market. Wholesalers are the trade's leading actors, they control the products obtained by the different towns in the area through the weekly souks,[8] first and foremost those of the Wilaya de Djelfa, ending up in part at the '*Qashabiya* Souk' in Messaâd. In this way the town's market is able to supply the country's various regions, whether the Centre, East, West, Pre-Saharan Zones or the South.

It is from these regions that secondary flows depart, on the one hand to wilayas and towns in each of the regions concerned, on the other hand to export countries. The products exported target first populations in neighbouring countries and the Middle East, and then the principal Arab and Muslim migrant populations in Europe.

Second, it seems that products of national origin also depart from neighbouring countries, as exports to Europe or the Middle East. In other words, the national product is appreciated and has a market abroad but is neither visible nor recognised due to the lack of alignment with modern marketing practices (labelling for example).

Third, it is essential to identify the 'nodes' of this network (value chain), which is relatively simple in terms of the interlocking and complementarity of the various routes or channels taken by the different types of *qashabiya*, as will be referred to below. In fact, what appears more complex is the way in which the relationships and practices of the different categories of actors are connected. The many interviews carried out led us to identify three essential nodes. The first node is the *marketing phase*, where the female weaver is faced with a lack of physical access to the market and so seeks the help of an 'intermediary' who acts as the interface with the commercial actor. It is this 'lock' that characterises so specifically this system of marketing and the network as a whole given the knock-on effect it has on the upstream production, within the household. The second node, which controls the flow of products collected to the country's regions and to export countries, is located at the *Messaad market* which, along with Deldoul, is the pole of this weaving activity, which consists of less than a dozen wholesalers who dominate the marketplace. The third node is that of the *exporting traders* who play the role of *exporters* and are based in the capital Algiers and Tebessa in the East, and Oran and above all Tlemcen, in the West (Figure 4.4). At the end of the value chain, whatever the route in which the *qashabiya* was channelled, we find the consumer.

Different flows by product depending on destination

As shown by the results of the two classification matrices below, the product flows are clearly different depending on the country of destination. The most obvious difference is the distinction between articles in camel hair (*wabar*) and the rest, the former largely flowing to the west and south. This applies to the entire range, the industrial (harrara or fibrame) products flowing less to the west, and for export, less to Morocco and Spain. The second matrix indicates that articles in wabar go mainly to the Maghreb countries and Emirates, those made of wool to Tunisia and Morocco (Table 4.1). The exported quantities elsewhere, including to migrants in Europe, are more limited.

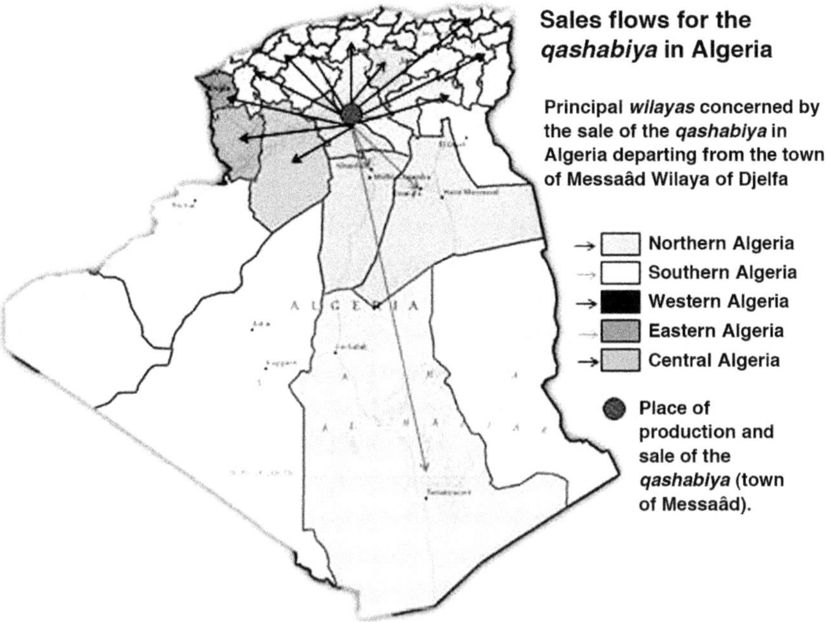

Figure 4.4 National and international flow: marketing flows of *qashabiya*

Figure 4.5 below shows that in terms of importance the flows are relatively similar to the centre, the west and the western border, while the east and the eastern border as well as the south exert less of a pull. The most important flows are towards the Moroccan borders and the center of Algeria, followed by the Tunisian borders and the east of Algeria.

The unequal valuing of labour and know-how and the unfair sharing of the value added to the detriment of the woman weaver

The matrix below, regarding the white *qashabiya*, shows that its manufacture involves 11 operations in total, providing a profit that reaches 13,500 DA (Algerian dinars; about 188 US dollars in 2012), see Table 4.2. Of this amount, the female weaver receives about 43 per cent, in second place, the tailor receives 22 per cent, while the local and souk traders together receive an equivalent amount. Also the female worker involved in subcontracting activities earns 1,050 DA (close to 15 US dollars), which means that women workers as a whole receive about 50 per cent of the total profit. The share of workers involved directly in *qashabiya* production, the women and couturier, equals 9,800 DA (136.10 US dollars), or 59 per cent of the total value. These figures reflect the share going to traders whose expenses are

Table 4.1 Classification and evaluation matrix for international *qashabiya* flows

Transit channels		*Transit Q from east to neighbouring and other countries*			*Transit Q from west to neighbouring and other countries*			*Transit Q to centre*	*Transit Q to south*	*Transit Q from west to European countries*	
Destination			*Export*			*Export*					
Range		*East*	*M*	*Tunisia + Libya and France*	*West*	*M*	*Morocco + Spain*	*Center*	*South*	*North*	*Export Europe France + Italy*
Camel	Adi	1	3	3	7	3	3	1	5	1	2
hair	Imported	2	2	4	2	2	2	2	4	2	1
	Aguiga	6	1	1	3	1	1	3	7	3	3
Wool	White Beidha	7			5		4	6	3	4	4
	Draa	3			4			7	6	8	
	Mixed										
	Lamkhalta										
	Colored/	4			8			4	2	5	
	El'Maqtoussa	5			1		5	5	1	6	
Industrial	*Harrara* or *Fibrame*	8	4	2	6	4		8	8	7	

Key: 1 = the most important flow towards foreign countries; 8 = the lowest flow; M = Messaâd

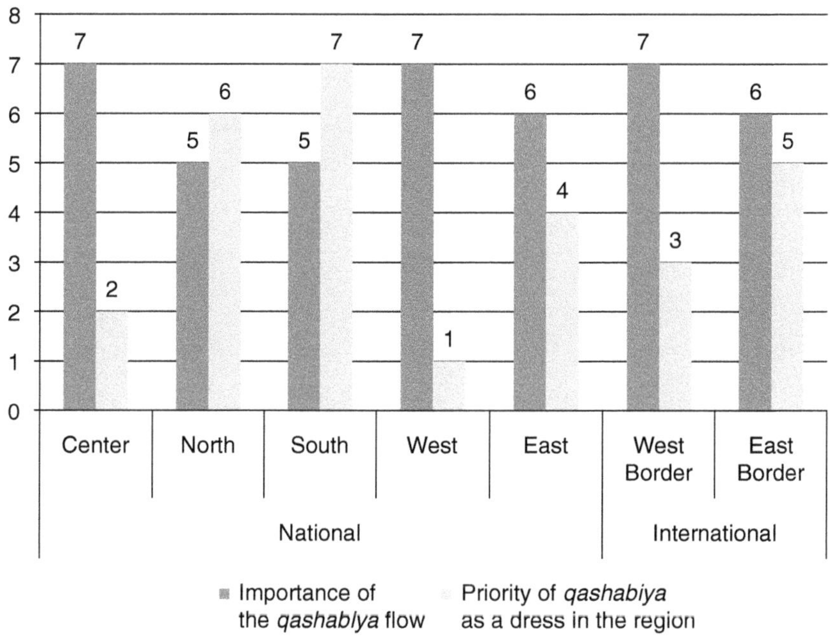

Figure 4.5 Qashabiya flows within Algeria

Key: 1 = low importance through to the highest importance 7

considerably fewer and have no significant investments. In most cases, with the exception of a few large wholesalers, these traders also escape tax levies, the transactions being effected largely in the informal economy. Also to be stressed is the supreme effort required by female artisans and subcontractors mainly for fiber processing, although very few of them sub-contract other women for weaving *qashabiya*.

The results show a 50 per cent income for women from *qashabiya* sales. The main constraint is their lack of physical market access where they can get a better deal, and also, importantly, a lack of women's cooperatives where they can bargain and get better deals for their products. At the moment, women's cooperatives are very limited in the area, and women have expressed their need for special shops or cooperatives where they can sell their products without significant loss.

The share of the added value that goes to women is of course dependent upon the number of operations they undertake as well as the marketing route adopted. The table below shows that the highest amount (94.44 US dollars) is obtained when the marketing is made without the intervention of a Dallal with an implication on all the production activities involved. At the other extreme, the profit drops to 59.02 US dollars when the production of *qashabiya* does not require all the processing operations (Table 4.3). The profit is quite similar for the other four scenarios.

Table 4.2 Expenses, sales, profits, utility and effort per category of actors and type of operations in the *qashabiya* chain (per *qashabiya*)

Actors	*Operations*	*Expenses (US$)*	*Sales (US$)*	*Profits (US$)*	*Utility* (%)*	*Effort** (%)*
Livestock producers (raw material)	1		2.91	2.91	1.55	10
Wool trader (raw material)	2	4.16		1.25	0.66	10
Female subcontractor for processing wool and camel hair/ Pre-weaving	0					
Weaving accessories	3	5.55		5.55	2.96	10
Female artisan/weaver	4, 5, 6, 7	9.72	111.11	94.44	50.00	40
Intermediary	0					
Trader (market)	8		118.05	13.88	7.40	10
Trader (shops)	9		145.83	27.77	14.81	10
Tailor (man)	10			41.66	22.22	10
Total	10			187.5	100.00	100

Note: 1 US$ = 72 DA (Algerian dinars) in 2012

Key: 1 = sale of raw material; 2 = sale from trader; 3 = accessories; 4 = cleaning, washing wool, combing; 5 = carding and spinning; 6 = mounting the loom; 7 = weaving; 8 = sale at the market; 9 = sale at small shops in the city; 10 = Sending the *qashabiya* for sewing to the tailor.
* Utility % = profit/total profit ×100
** Effort % = 1/12 ×100

Table 4.3 Women's income based on the number of operations and the type of marketing route taken by the product

Number of operations for the qashabiya *production**	*Scenario according to route type and number of operations*	*Profit (US$)*	*Utility (%)*	*Effort (%)***
1 in 12	Without the processing operations	59.02	31.48	8.33
3 in 11	With subcontractor	73.61	39.25	27.27
4 in 11	With weaving + intermediary	87.50	46.66	36.36
4 in 10	Without intermediary + all processing operations	94.44	50.00	40.00
2 in11	Two operations	79.86	42.59	18.18
3 in 10	With subcontractor + intermediary		43.00	30.00

* 1 = livestock producers, 2 = camel hair and wool trader; 3 = sub-contracting (including 3, 4, 5 for processing); 6 = trader accessories; 7 = mounting the loom, weaver; 8 = intermediary; 9 = weaver; 10 = market trader; 11 = trader in small shop; 12 = tailor

** = calculated as a percentage of the number of operations from the total (12)

The analysis included also the income shares from two types of *qashabiya* (Table 4.4) among the different actors. This share varies depending on the extent of the added value received by women weavers for the work they performed. In the case of the *qashabiya* made of Iraqi wabar (that involves four processing operations), the female artisan receives 18,000 DA (250 US dollars) out of a total profit of 40,500 DA, which represents less than half (44.44%). The finished *qashabiya* product is sold by women weavers for 25,000 DA and by the trader for 35,000 DA. The total profit margin taken by the market operators therefore reaches 10,000 DA or 40 per cent which would have been attributed to women had they been able to market their product directly with no intermediaries and other market actors. It should be noted that the finished product can in fact pass through the hands of several market actors, each one taking part of the total profit margin. Job opportunities in agro-pastoral areas – where industrial activities are yet to take off – livestock production activities and marketing in local markets are virtually the only accessible employment opportunities, hence the very high number of intermediaries (dallals). A committed policy to support job creation for youth, as well as the modernization of the red meat production value chain, with the creation of slaughterhouses, storage and packaging facilities for sheep production, has been initiated by public authorities as part of the Agricultural and Rural Renewal policy, and is now in progress. The result should be a considerable increase in job opportunities.

Power and domination relationships: losses, interests and respective profits of the different categories of actors involved in the qashabiya *chain*

An economic analysis of such phenomena can either be made on the basis of the quantitative determinants (indicators) such as prices based on supply and demand, time and capital, or can also take into account the power relationships of actors, and the various aspects linked to the personalities of the actors that determine their

Table 4.4 Estimation of benefits sharing: camel hair and wool *qashabiya*

Actors	*Camel hair revenues (US$)*	*Wabar Iraqi wool revenues (US$)*	*Income shares of camel hair (%)*	*Income shares of wool (%)*
Camel hair trader (M)	69.4	69.4	14	12
Accessories trader	13.9	13.9	3	2
Artisan/weaver (W)	250.0	333.3	69	44
Intermediary (M)	34.7	0.0		6
Retailer (M)	125.0	0.0		22
Artisan Tailor (M)	69.4	69.4	14	12
Total	562.5	486.1	100	100

Key: 1 US$ = 72 DA; M = men; W = women

Note: Figures for intermediary costs for camel hair are not available because most of the camel hair *qashabiyat* are sold either directly to the consumer or brought to the retailer by men of the household especially since only one *qashabiya* is ready at a time. The sale prices were roughly estimated by the respondents as being: 25,000 DA when sold directly by women to consumers and 35,000 DA when sold by a trader behaviour. The CLIP tool is strong for conducting investigations into these dimensions which are very often absent from studies. Figure 4.6presents the results of the exercise on the power and influence of different actors in the *qashabiya* market.

At the bottom of the scale is the woman who works on a subcontracted basis for a woman weaver, either processing camel hair and wool or weaving. The most influential actors are the men within the household, the wholesaler and the Dallal. Generally speaking, the various actors are perceived by the elder actors as legitimate and the practices in accordance with the 'natural order'. The young people, on the other hand, do not accept the current order and demand fairness in the transactions. However, this position does not encompass an affirmation by men and women artisans to make an effort and reach a common position for defending their interests.

The factors that hinder marketing: the wool made qashabiya

The results set out below show to what extent *qashabiya* marketing faces constraints. The force field tool was used to try to highlight and understand the interplay of the various factors. Figure 4.6 shows that three of the five hindering factors identified are of equivalent weight: the strategies of traders to obtain a

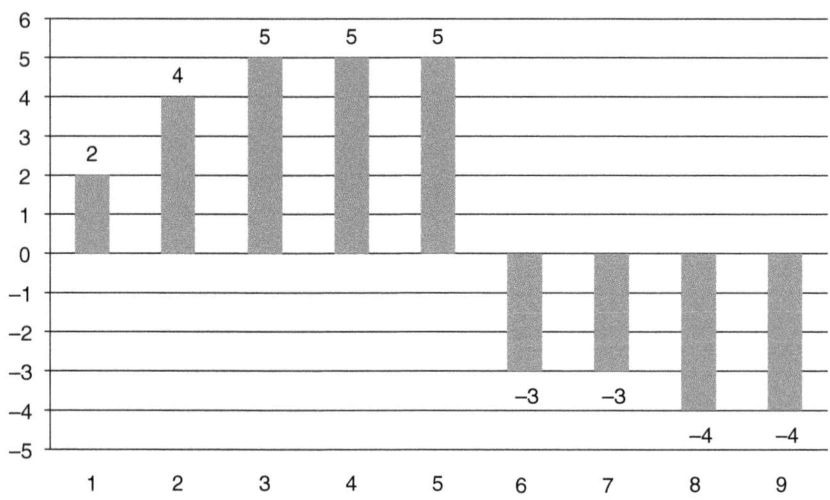

Figure 4.6 Force field – alleviating and constraining factors for *qashabiya* marketing

Key:
1 = Price volatility of raw materials related to the seasons and different festive occasions
2 = Multitudes of stakeholders in the marketing of *qashabiya* (intermediary, tailor, traders)
3 = Strategies of traders' monopoly of the market
4 = Selling directly to consumers without passing by the market
5 = Import of raw material (Iraqi fiber) and emergence of competition with local camel hair
6 = Use of websites and the local media
7 = Participation in exhibitions and fairs
8 = Performing a traditional Souk (100 stores) program of the President of the Republic
9 = Programming a national day for weaving and wool products in the region Messaâd

monopoly, the indirect method of sale, and the emergence of imports of an innovative raw material (the Iraqi wabar) that has entered into keen competition with the local raw material (the local wabar) traditionally used. The factors that, on the contrary, have an alleviating impact, are when sales are made directly to consumers without passing through the market, with an approximately equivalent weight. Some premises are made available to women and men professionals, and there has been a recent initiation of fairs and exhibitions for local products, and quality promotion, in particular through labelling. However, women are still not allowed access to participate in the market place. Therefore, female-managed premises could constitute a substitute and provide space to sell their products at the best prices, because they would be able to bargain and set their prices according to the costs of raw material in addition to their multiple processing and weaving efforts. This idea was expressed by most women interviewed, and will constitute a starting mechanism for the creation and multiplication of women's associations for the promotion of local products. A women-run market place supported by women's organizations can overcome the effects of traditions that do not allow women to participate in the male dominated markets and can empower women through better opportunities to bargain over their products.

Diagnosis and perspectives: generated by the SWOT tool

The SWOT tool was used to identify the essential elements of the value-chain diagnosis, which are detailed by order of importance:

Opportunities

There is a clear demand for the local product, *qashabiya*, of different types, rehabilitated and strengthened by recent changes in the social values (return to rural values, religious sentiment). It also has an important export potential due to international demand. Existence of large scale programs and projects initiated by the government as part of sectorial policies to rehabilitate and add value to local products are indication of national support for local products. This product could form a niche that could attract young people in agro-pastoral regions in particular (during the period needed to initiate and realise projects in other sectors and to create jobs). It also forms a niche to provide significant additional income without crippling budgetary resources, with a potentially significant and immediate impact on reducing vulnerability and poverty in the countryside. This positive outlook is a reflection of the favourable social context created by the strongly expressed political will for the economic and social–political emancipation of women, in particular in rural areas (also seen in the legislation considered to give women one third of places on electoral lists in October 2011).

Strengths

The strong points of the *qashabiya* subsector are:

1 The know-how for this product has been recognized as a national cultural heritage that should be preserved.
2 The availability and demonstrated desire of the elderly men and women who possess this knowledge and know-how to ensure this knowledge and know-how are preserved and passed on to the youth.
3 The reorientation of and changes to the various sectorial policies concerned, with measures and programs already launched since 2000.
4 The strong support from national policy:
   a National Policy for Agricultural and Rural Renewal: inclusion of the preservation and valuation of the intangible heritage and artisan activities in the PPDRI (National Strategy for Rural Development and Integrated Rural Development Proximity Projects); for example for labelling of local products and the impetus given to exports
   b The National Policy for the Rehabilitation and Development of Artisan Activities: programs for training, financial support, counselling, capacity strengthening of the profession (regional chambers, associations, support for the creation of service cooperatives), and the mobilisation of foreign expertise (cooperation projects)
   c The National Trade Policy: creation of a national markets network at local, regional and national level; actions to regulate markets and absorb the informal market
   d The National Vocation Training Policy: aimed at providing training that meets local and regional needs so as to capitalise on the specific strengths of rural areas
   e The National Policy for Scientific Research and Technological Development, aimed at guaranteeing scientific and technological support for the economic actors and institutions concerned
   f Finally, the Local Development Policy that aims to strengthen links between institutions (central government, regional and local authorities, professions, universities and research institutes, NGOs, etc.). Current action pursued by the CNES (National Economic and Social Council) on the instructions of the President of the Republic, studies and debates in the field to take stock of the situation on the ground and participative diagnosis of local development, with identification of useful recommendations for its strengthening.

Weaknesses and threats

There are numerous weaknesses in the system. Among these are:

1 Weak ability to organise and act cohesively, leading to a lack of strong representative partner from the government and the various institutions concerned (local and regional authorities APW-AOC, etc.), and at the various levels (local, wilaya, per large region: steppe and national).
2 Slow progress in building a recognised and operational interface between the

actors and government, making it necessary to seek the causes of the low impact of the actions initiated (approaches? tools?).

3 Insufficient coordination and lack of synergy among method of intervention of various government structures and various other institutions – including training, research and popularising institutes and centres, and the associations – although still evolving.
4 Limited impact of facilitating access of potential beneficiaries to information about the support program initiated by government.
5 Insufficient extension support to male and female artisans, reaching only a limited fraction of potential beneficiaries.
6 Insufficient monitoring and evaluation programs and projects that support artisan activities and low participation in evaluation.
7 Cultural constraints related to the values and traditions and therefore behaviour of the actors, which impede women's participation in the market place and other institutions, remain very pertinent and seem to be insufficiently recognised.
8 Predominance of the informal character of the market and the entire value chain in general, lack of transparency, traceability, reliable statistical data and significant application of regulations (taxation).
9 High risk of continuing and accelerating the loss of artisan knowledge and know-how relating to the *qashabiya* types if the unequal sharing of benefits to the detriment of women persists, which discourages young people from choosing these activities.
10 High risk of an increase to the negative impact caused by the import of technological innovations through raw materials that affect the quality of the *qashabiya* product and loss of market share for products that respect traditions.
11 Risk of insufficient job availability to generate a regular income for households, in particular for male and female youths, and for the smallest crop and agro-pastoral farms in remote rural areas (leading to poverty, food insecurity, lack of social cohesion, etc.).
12 Lack of recognition of artisans in their rightful place as essential levers in promoting tourism.

Conclusion

The results of this research clarified the key components of the present situation, thus the opportunities and the strengths as well as the weaknesses and the threats that weigh heavily on the future of the *qashabiya* market and along with it the related ancestral knowledge and know-how, with resulting consequences for members of the rural communities. It emerged clearly that the rural woman, the principal actor in creating the added value, is disadvantaged and confined to a social straitjacket that denies women the necessary access to market and the autonomy that allows them to assume a full role as economic actors formulating projects and taking consequent strategic and management decisions. This situation has a high social cost, in terms both of unrealised jobs and income, scope for progress in food security, reduction in the insecurity and poverty of households and delays in the recovery and sustainable development of the rural areas concerned.

It seems clear that if the key constraint is to be overcome to permit the success of the programs and projects already initiated under genuinely committed sectorial policy on the part of government, which has allocated the budgets needed to meet the social needs identified, then resolute action is needed by government to promote and develop in a sustainable manner this pivotal artisan product for agro-pastoral regions. This resolute action must be coordinated with the social actors to accelerate the pace of social change (regarding place and role of women), adjust the relationships between the different actors, and permit a fairer sharing of the added value created. The key axis remains a strengthening of the capacities of the various actors, concerning all categories and the coherent group that these must forge with the help of government.

The possible solutions therefore relate more to the qualitative aspects – the functioning of the economic activities at different levels, including within the household, and the way in which these activities take place and slot into regional and territorial governance – rather than the technical aspects that seem more accessible given that the human capacities exist and are waiting to be mobilised (universities, training and research centres).

Notes

1 L'École Nationale d'Agriculture, El-Harrach, Algiers.
2 L'Institut National de la Recherche Agronomique (INRA), Algiers.
3 The International Center for Agricultural Research in the Dry Areas (ICARDA), Amman.
4 L'Institut National de la Recherche Agronomique (INRA), Djelfa.
5 Institut Technique Moyen d'Agronomie (ITMA) Djelfa.
6 Governorate.
7 Dallal is an intermediary.
8 Local market.

References

Ageron, C. R. (1971) *Les Algériens musulmans et la France 1871–1919*, 2 volumes, PUF, Paris.

Benachenhou, A. (1980) Formation du sous-développement, Essai sur les limites du Sous-Développemnt en Algérie, OPU, Algiers, p. 394.

Benaoum, A. (1978) Recherche, Etude de classes sociales en milieu pastoral, in: Revue Débats et critiques (now 'La revue du CENEAP'), numero 1, CENEAP, Algiers.

Bessaoud, O. (2008) L'agriculture et la paysannerie en Algérie, Les grands handicaps, in N. Benghabrit-Remaoun and M. Haddab (eds) *L'Algérie 50 ans après, Etat des Savoirs en Sciences Sociales et Humaines, 1954–2004*, CRASC, Oran, pp. 359–84.

Bessaoud, O., Chassany, J.P., Abdelhakim, T. and Nawar, N. (2005) Le développement rural durable en Méditerranée, in Agri-Med, 2005 annual report, CIHEAM, pp. 1–65.

Bessaoud, O., Bourbouze, A., Campagne, P. and Le Cotty, T. (2000) Problématique du développement rural des zones sèches de la région MENA (Moyen Orient et Afrique du Nord), contribution to updating of the report 'From vision to action for rural development', compiled by the World Bank, Ministry for Cooperation, France, World Bank, Washington, December 2000.

Boukhobza, M. (1982) *L'agro-pastoralisme traditionnel en Algérie, de l'ordre tribal au désordre colonial*, OPU, Algiers.

Bourdieu, P. and Sayad, A. (1964) *Le déracinement, la crise de l'agriculture algérienne en Algérie*, De minuit, Paris, p. 224.

CENEAP/FNUAP (2001) *La participation de la femme au développement économique et social*, CENEAP, Algiers.

Chaulet, C. (1997) Agriculture familiale et modèles familiaux en Méditerranée, réflexion à partir du cas Algérie, in A. Abaab, P. Campagne, M. Elloumi, A. Fragata and C. Chaulet (eds) *La terre, les frères et l'argent, Thèse d'Etat (Sociologie)*, OPU, Algiers.

Chouaki, S., Cherfaoui, M.L., Aït Hamlat, N., Moussaoui, M., El Mourid, M., Abdelali-Martini, M. and Buerli, M. (2010) Renforcement des moyens d'existence des ménages et développement rural: Cas de quatre villages du massif des Bibans (Commune d'Ighil Ali – Wilaya de Bejaia), INRAA/ ICARDA/SDC project, INRAA, p. 128.

Cherfaoui, M.L. (2009) Le Plan National de Développement Agricole et Rural: Vision, objectifs stratégiques, démarche, dispositif de mise en œuvre – Eléments de doctrine, INRAA, p. 44.

CNES (n.d.) La conjoncture économique (bi-annual), Algiers.

Colonna, F. (1987) *Savants paysans, Eléments d'histoire sociale sur l'Algérie rurale*, OPU, Algiers.

Cote, M. (1996) *Pays, paysages, paysans d'Algérie*, CNRS, Paris, 1996, republished by MEDIA-PLUS, Constantine, 2008.

Despois, J. (1964) *L'Afrique blanche: l'Afrique du Nord*, PUF, Paris.

Despois, J., and Raynal, R. (1967) *Géographie de l'Afrique du Nord-Ouest*, Payot, Paris.

El Kenz, A. (ed.) (1991) *L'Algérie et la modernité*, Karthala-Codesria, Paris.

Ministry of Foreign Affairs (MAE) (2009) Communiqué Audition ministre chargé de l'Artisanat par SE Mr le Président de la République, MAE/RADP and national press.

Minister for Rural Development (MADR) (2006) Rural Renewal, August 2006, p. 211.

Deputy Minister for rural Development (MADR) (2004) National Strategy for Sustainable Rural Development, project, July 2004, p. 44.

Ministry of Trade (CNIS) (2011) Export balance, period January 2011/September 2011.

Ministry for Small and Medium-Sized Enterprises and Artisanat (2008) Report on the 6th Rencontre Internationale de l'Artisanat Traditionnel, Développement de l'artisanat et promotion des échanges extérieurs, held in Algiers on 30 May 2008.

Toualbi, N. (2001) *L'identité au Maghreb, L'errance*, 2nd edn, Casbah, Algiers, p. 270.

5 Occupational health and prospects of empowerment for female weavers

Evidence from a home-based activity in the informal sector in Djelfa, Algeria

Malika Abdelali-Martini[1]*, Abdelmadjid Benmebarek*[2] *and Amel Meguellati-Kanoun*[3]

Introduction

From the perspective of medical doctors, occupational health is routinely explored through sex as an explanatory variable and to increase emphasis on the mechanisms involved in any sex or gender differences (Messinga and Stellmanb 2006). This was found in the bakery sector where specific respiratory allergies and asthma were clearly demonstrated as a result of constant exposure to flour and other ingredients (Houba *et al.* 1998). In a study comparing compensable work-related injuries and illnesses between females and males across all major industrial sectors and occupations have found that the overall injury or illness rate was significantly lower in females than males (5.5 vs. 11.5 per 100 employees), a trend that extended to all major industrial classes with the exception of service and agricultural sectors. The distribution of types of injury or illness varied by gender, occupation, and industry with significantly higher risk of carpal tunnel syndrome, burn, sprain, and fracture in females compared to males (Islam *et al.* 2001). In a study conducted in the Iranian carpet industry to determine the prevalence of upper limb problems associated with hand tools design, and focused on designing or redesigning the weaving comb, knife and scissors as the most common tools in carpet weaving operation Motamedzadea *et al.* (2007) found that the new ergonomically designed weaving hand tools were found to be effective and acceptable to the carpet weavers, and that ergonomic improvements in the carpet hand tools may promote health and safety among carpet weavers.

In the governorate of Djelfa in the south of Algeria, poor women earn livelihoods by processing camel hair and raw wool material and weaving a national garment used as traditional coat and worn for prestigious national occasions. It is also often a favored gift for and given by the highest levels of the national hierarchy to eminent guests. However, while this activity could constitute an important means for women's empowerment, it is greatly affecting their health mainly because it is a problem largely ignored by health and development agencies and being a home-based activity, it is not covered by national insurance schemes.

The informal weaving sector in Algeria is generating important income for rural and urban communities, but is also challenging women's health: women are responsible for about 90 per cent of the activities in this sector. Traditional weaving is transforming by adopting more easy-to-use tools and mechanization as awareness grows about occupational safety and rights within society. This paper provides evidence from South Algeria addressing a specific, home-based, occupational health issue related to the production of a national symbolic garment (*qashabiya*). With globalization challenges and the economic transition that carries modern industries, the State needs to be alerted of this problem and should include in its development strategies proper legislation to accommodate women home weavers' health and an overall inclusive growth. This is also important to save a traditional national gown and its economic role.

The challenge is how to ease home-based women's work and facilitate a production process of better resilience (or less vulnerability)? Therefore, the main research questions addressed in this research are: 1) What types of ailment are associated with the different women's processing activities in the production of *qashabiya*? 2) How have these ailments been addressed at local levels? and 3) How do improvements to their working conditions empower them, through issuing proper policies, to overcome this important health issue?

We hypothesized that women home-weavers' health is affected by a number of occupational ailments that are not covered by the national insurance scheme, have not been properly addressed, and therefore, not treated properly. This could also be because home-weaving is not recognized as official private or state employment.

Methodology

In Djelfa governorate (described in Chapter 2 of this book), we used a methodology mainly based on gender analysis and qualitative tools selected from Social Analysis Systems (SAS[2]) (Chevalier and Buckles 2008) to address specific issues related to gender relationships and the use of natural resources in marginal areas, and more specifically health issues related to women weavers.

We used qualitative methods to identify problems that people, including women, face, the actors or stakeholders affected by a situation or with a capacity to intervene, and the options for action. From this we have used the 'problem tree' participatory technique to assess the two levels of causes and effects of occupational health issues and the local treatments administered, and the 'force field analysis' to identify the factors that cause or perpetuate the identified problems, and those that counteract/mitigate them.

Focus group discussions were held with women from different social groups and of different ages, involving some in-depth interviews to deepen our understanding of these problems. We also visited 15 specialized and general doctors in Djelfa town and its surrounding area to find out whether the *qashabiya*-related ailments were well understood as related to a home-based activity, and treated accordingly. During the implementation of this project, the research team visited a number of associations and encouraged more focused discussions on potential relationships between occupations and ailments of women.

Conceptual framework for empowerment

In order to better understand and analyze health issues of home-based female weavers, we use a conceptual framework to guide our analysis of women's empowerment in rural Algeria inspired by Mayoux (2003) and Luttrell *et al.* (2009). According to this framework, the process of a woman's empowerment can be understood through free and informed life choices that are only possible within an environment of equal opportunities, resources and power. It is shaped by several dimensions representing:

1 the power from within or perceptual change;
2 power to or changes in capacities and material strength;
3 power over or relational change; and
4 power with or organizational strength and participation.

Others have also considered empowerment through four dimensions: self-esteem, participation in household decision-making, freedom of mobility and control of material resources (Mahmud *et al.* 2012). In this context, a woman's freedom of mobility, control of resources and participation in decision-making change over her life cycle and rise with age. In the case of Djelfa in Algeria, where social norms constitute a barrier to women's development, the analysis of empowerment through the different aspects listed above holds the potential of better targeting development options. In any case, income-generating activities successfully empower the individual performing the activity as well as his or her entire family (Becker 2012 in Alkire *et al.* 2012; Charlier *et al.* 2007). Others refer to these different aspects of empowerment as the assets, knowledge and know-how (power to), will (internal power), and capacity (internal power and power with) (Charlier[4] *et al.* 2007).

Despite diversities among countries, women's freedom in the Arab States and beyond changes gradually as women get married, have children and get older. They are then allowed to move more freely beyond the household boundaries, participate in decision making and have control over some resources. To some extent, this is more the case in poor households. In wealthy households, where women do not perform any income-generating activities, women's empowerment is generally limited to their control over some resources within the household's boundaries where they take major decisions especially when their sons get married and share the same house.[5] This differs largely between urban and rural areas as well as between high and low income households, in addition it is related to education levels of women and men. Furthermore, the degree to which customs and traditions impact on decisions and behavior is another key dimension of these differences. Exposure to collective work and market has the potential to empower women in many ways. Information and new ideas as well as capacity building and exposure to the outside world have the potential of enhancing self-esteem and opening new avenues for empowerment and improvement in working conditions.

Results

Profile of women weavers who process raw material and undertake weaving activities

Women weavers processing wool and camel hair are mainly from poor households who work in haste making one *qashabiya* at a time because they use the proceeds from the sale of one *qashabiya* to buy the raw material for producing the next one, as they do not have a capital that allows them to produce many items before deciding to sell when prices are satisfactory. This pushes women to sell their products at low prices. The situation puts women in a weak position, especially as they do not own property, have no credit, are generally absent from cooperatives, and their fall-back position is weak. They also lack access to information about governmental strategies, which break women's roles from the beginning. Overall, the raw material processing and weaving activities performed under precarious conditions using traditional equipment puts their health at high risk, in particular as they do not benefit from any insurance coverage.

In general all activities could be performed either individually or collectively. However, in our sample, women weavers work either individually or collectively depending on the type of *qashabiya*, and on availability of women both within the household and from the neighborhood when they agree to exchange labor between themselves (Table 5.1). Individual activities comprise the first stages of the production process, from washing wool and camel hair up to combing, and when necessary dying the wool. Most of the other activities are performed collectively either for socializing, sharing 'what is up' and local stories, fostering a sense of belonging and building mutual support

Women weavers of different ages face a number of health and safety risks, caused by long working hours, cramped workplaces, and malnutrition due to poverty, poor healthcare facilities and lack of recreational opportunities.

Ailments associated with the different women's processing and weaving activities

As a largely informal weaving sector, a number of hazards affect (fraught) *qashabiya* pre-weaving and weaving activities. These are exacerbated by poverty and small weaving spaces, with poor natural lighting and poor ventilation. Old equipment is still used in the whole area, with almost no modification. It is passed on as it is from one generation to another.

Qashabiya weavers and raw material processors suffer greatly from a number of serious ailments (Table 5.2) which include joint ailments such as skeletal deformations, allergies, injuries/wounds, eyesight disorders, cephalous and stress affections and other toxic hazards due to the dye. The type of ailment depends on the nature of activities, type of *qashabiya* produced, age, and the person's overall health conditions. Our research has identified them as linked to their weaving related activities, but they may also be due to their precarious living conditions.

Table 5.1 Activities by type of fiber used to make *qashabiya*

Activities	*Camel hair*			*Wool*			*Industrial fiber*		
	*Aguiga**	*Adi*	*M'khalta*	*Iraki*	*Daraa*	*Beidha***	*M'khalta*	*Maqtousa*	*Fibrame*
Washing (Gsil)					S	S	S	S	
Echarpage (Bachm/thanthir)	S	S	S		S	S	S	S	
Mixing (Tekhlat)									
Combing (M'shit)					S	S	S	S	
Carding (Tkardiche)	C	C	C	C	C	C	C	C	
Spinning (Ghzil)	C	C	C	C	C	C	C	C	
Dying (Qatssa)					S	S	S	S	
Mounting loom (Sédoua)	C	C	C	C	C	C	C	C	C
Tissage (Nsij)	C	C	C	C	C	C	C	C	C

Key: C = Collective activities; S = Specific activity – i.e. activity performed individually;
**Aguiga* is the specialty of Mesaad; ***Beidha* is the specialty of Zaccar and Znina villages/towns.

Joint ailments and other related discomforts

Joint ailments affect women of 45 years and older. These women are also in charge of the domestic chores and have gone through many child deliveries that have contributed to weakening their health. Weavers occupy specific positions: squatting on the old type of loom for weaving, and paying due attention and leaning forward, which have led over time to some serious skeletal problems, especially for the elders. These are often compounded by the type of activity they perform in processing wool and camel hair, as well as nutritional deficiencies associated with poverty. Among young women, the hands become deformed (genu valgum), or a crippling arthritis of the knee may develop during winter, especially during wool spinning by rolling the spinning tool over the leg while the knee and leg remain uncovered. The constriction of the pelvis that sometimes occurs in females may make it necessary for them to have a Cesarean delivery when giving birth. Lateral curvature of the spinal column (scoliosis) and back pain are also common ailments. These ailments were discussed with and confirmed by general and specialized medical doctors in Djelfa (Table 5.3).

Table 5.2 Types of ailments affecting women by activity

Washing Ghsil	*Teasing or picking/* Bachm thanthir	*Washing* Ghsil		*Teasing or picking/* Bachm/ thanthir		*Combing* M'shit + Tekhlat		*Carding and roving* Tkardiche		*Spinning* L'aghzil		*Dying* Qatssa		*Mounting the loom* Sédoua		*Threading/* *Teniar*		*Weaving* Nsij	
		I	*C*	*I*	*C*	*I*	*C*	*I*	*C*	*I*	*C*	*C*	*I*	*I*	*C*	*I*	*C*	*I*	*C*
Joint diseases	Finger	2		3	2	3	2			3	3							3	3
	Wrist	2		2	2	2	2	3	3	3	3							3	3
Arthritis/ deformations	Swollen hands/ clocks	*2*		2	3	2	3	3	3									3	3
genu valgum	Shoulder	3		3	3	3	3	3	3					3	3	3	3	3	3
	Neck	3		3	3	3	3	3	3									3	
	Arm	*3*		2	3	2	3	3	3	3	3							3	
	Ankle	2		1	2	1	2												
	Knee	3		2	3	2	3	3	3	3	3							3	3
	Back foot									3	3					3	3		
	Back	3		3	3	3	3	3	3	3	3			3				3	3
Allergies	Skin red/ eye watering			3	3	3	3	3	3										
	Cough	3		1	3	1	3	2	2							2			
	Sneezing	3		2	3	2	3	2	2										
	Asthma	3		2	3	2	3					3	3						
	Redness + skin inflammations eczema	3		1	3	1	3			3	3	3	3						
	Irritations burn in the throat	3		2	3	2	3												

Table 5.2 continued

Washing Ghsil	*Teasing or picking/* Bachm thanthir	*Washing* Ghsil		*Teasing or picking/* Bachm/ thanthir		*Combing* M'shit + Tekhlat		*Carding and roving* Tkardiche		*Spinning* Ghzil		*Dying* Qatssa		*Mounting the loom* Sédoua		*Threading/ Teniar*		*Weaving* Nsij	
		I	*C*	*I*	*C*	*I*	*C*	*I*	*C*	*I*	*C*	*C*	*I*	*I*	*C*	*I*	*C*	*I*	*C*
Allergies (cont)	Mouth gum ulcer	2		1	2	1	2												
Injuries	Cracks, finger whitlow	3		2	3	2	3					2	2			3		3	3
Ophthalmologic	Low vision	2		3	2	3	2	3	3							2	2	1	3
Cephalous	Headache/ dizziness													3	3	3	3	3	3

Key 0 = No ailments; 1= low; 2 = average; 3 = important ailment; I = individual; C = collective

Notes:

1 Combing + mixing (*tekhlat*) = similar activity

2 There is no specific constraint related to mounting the loom (sedoua) for camel hair *qashabiya*, whereas for that made of wool and fibrame fibers, the operation is more difficult because for the first, wool needs to be pulled tight and for the second, fibrame fiber needs to remain loose. Only experienced women can do the different types of *sedoua*.

3 The operation of tay (folding the woven part of the *qashabiya*) often causes headache specifically for the *qashabiya* made of *Wabri*, and eye-related disorders when working with fibrame fiber. The *hak* (softening the finished product) performed on wool *qashabiya* causes sneezing and coughing.

4 Joint diseases, ankle problems and backache and other related discomforts appear mainly at the age of 45–50 years.

5 Allergies include eczema in children's faces and the elderly during the cleaning process.

6 Stress is mainly observed when work is performed under tight deadlines, then in cases when the *qashabiya* is not sold quickly and also in case women do not have enough money to buy raw materials.

Table 5.3 Main ailments and diseases that affect female weavers and male tailors in Djelfa

Orthopedic/joint diseases		*Allergists*		*Dermatologists*		*Pulmonologists*		*Ophthalmologists*	
Osteo-arthritis	W+M	Rhinitis	W	Eczema	W	Bronchitis	W	Narrowing of the retina	W+M
Sprains	W+M	Sinusitis	W	Abscess or cyst	W+M	Pneumonia	W	Myopia	W+M
		Pharyngitis	W	Whitlow	W+M	Respiratory failure	W		
		Laryngitis	W						
		Asthma	W						
		Conjunctivitis	W						
Rheumatism	W+M								
Joint inflammation	W+M								
Tendonitis	W+M								
Herniated disc	W+M								
Sciatic nerves	W+M								

Key: W: Woman weaver, M: Man Tailor

The main health and safety hazards faced by *qashabiya*-weaving women as observed in the study areas are: musculoskeletal problems due to an odd squatting position and continuous attention required during work; vision disorders due to dust, inadequate lighting, combined with close attention required for work; hand and finger deformation and wrist pain, cracks and finger whitlow, swollen hands with clocks; pain on shoulders, neck, arms, ankles, knees, soles, and backache; night pain due to manipulation of small threads of wool and camel hair. Women also suffer from bladder and urinary incontinence because they retain urine while sitting for long hours.

Women living in towns are less exposed to risks of bad working conditions such as type of habitat, exposure to cold, heating system (wood/fuel). Those living in modern buildings are less exposed to cold, because flats are closed and better heated. The floor remains hot even from the heating system of neighbors. However, small rooms in flats are not suitable for the Arabic traditional loom. Women suffer more from sore joints by manipulating thin thread. The constant pressing on the *carders*[6] with fingers while carding, and the constant and repetitive movements and threading while spinning have resulted in seriously swollen finger joints, arthritis and neuralgia causing finger disabilities accompanied with night pains, as also confirmed by medical doctors.

Allergies

Most types of allergies encountered in the sample were due to cleaning wool, and less to cleaning camel hair. Women producers of *qashabiya* beidha (white *qashabiya*) suffer more from these allergies because the white *qashabiya* is made exclusively of wool. Women producers of aguiga type (from young camel hair) suffer more from pain in their fingers due to the necessity of working with thin threads compared to the other types of *qashabiya*. There is no need to wash camel hair while this operation is a long process in the case of wool. Skin and respiratory allergies are experienced by women and children due to inhalation of wool dust during processing, and also young children constantly present next to working mothers suffer from these allergies especially on their faces. These respiratory problems, including laryngitis, which might also be aggravated by some chemical residues,[7] were reported by respondents. Chemical hazards were also reported caused by inhalation of wool that contains dyes and hazardous substances. Allergies include red eye/eye watering, coughing, sneezing, asthma, redness and skin inflammations, eczema, irritations, burn in the throat, ulcer of the mouth gum. Furthermore, Magatos[8] product is used on camels for scabies and chips. Used on camels for skin disease, it is transferred through camel hair causing allergies on women's hands. Others in different areas use Gresil[9] instead to cure camels, a product usually used to clean bathrooms and floors. Two types of Magatos exist: the traditional, and the veterinary. It is also used to combat lices and scabies. It seems that it may have caused goiter as reported by women.

Ophthalmologic and cephalous injuries

Vision disorder and eyestrain were reported by women weavers and wool and camel hair processors. Dust, in addition to the constant close focus on the thin threads under low light, may be at the origin of vision disorders. Weaving takes place most of the time during the day using daylight, but the weaving spaces have no windows and daylight is insufficient. Women weavers suffer also from a number of ophthalmologic discomforts such as low vision, and cephalous problems such as headache and dizziness due to the high concentration required for the job. Many women also weave by night – some of them do not have electricity – and must concentrate to reach an excellent quality of weaving that remains more visible on white than on dark *qashabiya*.

Stress

The high degree of skills and constant attention to detail over long hours are potent psychosocial stressors, which may be compounded by other household responsibilities and children's needs at home. Women home-weavers reported that they developed nervous behavior and stress because of a lack of social entertainment that is essential for emotional balance, and constantly working under the deadline of a potential buyer.

Ailments by type of **qashabiya** *and their intensity*

The ailments described and their intensities depend also on the type of *qashabiya* produced and the number of processing activities it requires. By order of importance, camel hair, then wool are the most difficult to perform. They require a large number of processing activities before weaving (Table 5.4). Weaving industrial fibers to make *qashabiya* is less laborious for women, but of course the revenue from synthetic *qashabiya* is very low compared to the others.

The intensity of ailments shown in Figure 5.1 indicates *fibrame* as an industrial fiber that does not require any processing; therefore the ailments were mainly due among other reasons to weaving-related activities in terms of bad position and precarious types of equipment. Figure 5.2 generated from the aggregate ailment effects indicates that all wool *qashabiya* types: daraa (natural brown wool), mixed, dyed and white (62, 61, 59 and 53) are the most detrimental to women's health. Camel hair (wabri) comes in at second place. Although it is very difficult to process and weave, this type of *qashabiya* is made by a limited number of women because the raw material is expensive and related expertise rare.

Health issues related to the production of *qashabiya* are more prevalent among women living in rural areas compared to those living in towns and performing the same operations. Although the reasons are multiple, the research indicates that some of them might be due to the type of housing occupied by the interviewees. Given the type of housing in multiple floor buildings in urban areas, with small rooms, women use smaller metallic looms instead of the larger, wooden ones actually used

Table 5.4 Types of ailments by type of *qashabiya* and fiber

Types	Qashabiya	*Types of ailments*																				
		Joint diseases					*Allergies*					*Eyesight*					*Stress*					*Grand*
		1	*2*	*3*	*4*	*Tot*	*1*	*2*	*3*	*4*	*Tot*	*1*	*2*	*3*	*4*	*Tot*	*1*	*2*	*3*	*4*	*Tot*	*Tot*
Wabri *Camel hair*	Aguiga	3	2	1	4	**10**	4	2	1	2	**9**	4	2	1	3	**10**	3	4	3	4	**14**	**43**
	Adi	3	2	2	2	**9**	5	4	1	4	**14**	4	2	3	3	**12**	4	3	4	2	**13**	**48**
	Mkhalta	4	4	2	3	**13**	5	4	1	3	**13**	5	2	2	2	**11**	3	2	5	1	**11**	**48**
	Irak	1	4	3	3	**11**	1	1	3	1	**6**	1	3	3	3	**10**	2	4	5	3	**14**	**41**
Wool	Beidha	5	5	4	5	**19**	4	5	1	2	**12**	4	5	3	3	**15**	3	4	5	1	**13**	**59**
	Daraa	5	5	3	5	**18**	5	5	1	3	**14**	4	4	4	2	**14**	4	5	5	2	**16**	**62**
	Mkhalta	5	5	3	4	**17**	5	5	1	3	**14**	5	4	3	2	**14**	4	5	5	2	**16**	**61**
	Maqtousa	4	5	4	4	**17**	5	5	1	2	**13**	3	3	5	1	**12**	3	2	5	1	**11**	**53**
Industrial fiber	Fibrame	–	–	4	4	**8**	–	–	2	–	**2**	–	–	4	2	**6**	–	–	5	–	**5**	**21**

Key:
1 = Cleaning
2 = Raw material preparation: carding, spinning, etc…
3 = Loom frame preparation (trame) – sedoua + teniar
4 = Weaving (*dag* + *hak*)
5 = Among others, concentration for mounting the loom performed only by experienced women.
Scores: 1 = lowest through 5 the highest

Notes:
1 The mode reported in the table has been calculated based on detailed information reported on tables during interviews.
2 Loom frame mounting (*sedoua)* causes stress because the success and quality of the *qashabiya* greatly depend on this operation. It needs to be well adjusted, otherwise the whole *qashabiya* quality is affected and its price reduced.
3 Weaving: *dag* is the most constraining activity. *Teniar* (threads alignment on the loom) is done once after mounting the loom.
4 The cleaning stage of raw material is determinant of the product's quality.
5 *Dag* and *hak* (weaving and or softening the woven part) are continuous activities.
6 Spinning Irak wool is constraining because of dust and spines contained in it.
7 Stress is mainly experienced during the mounting of the loom that is critical for the production of a good quality of *qashabiya*, and the efforts made to respect the delays in the delivery of final products (classification 1 the lowest through to 5 the highest).

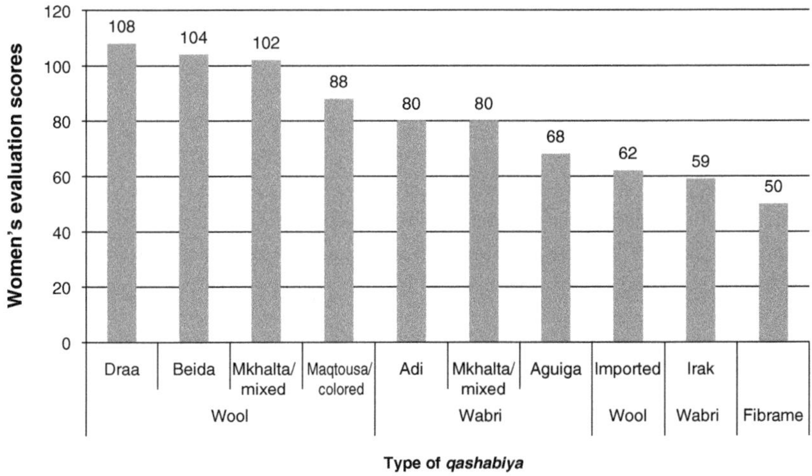

Figure 5.1 Intensity of ailments by type of *qashabiya*

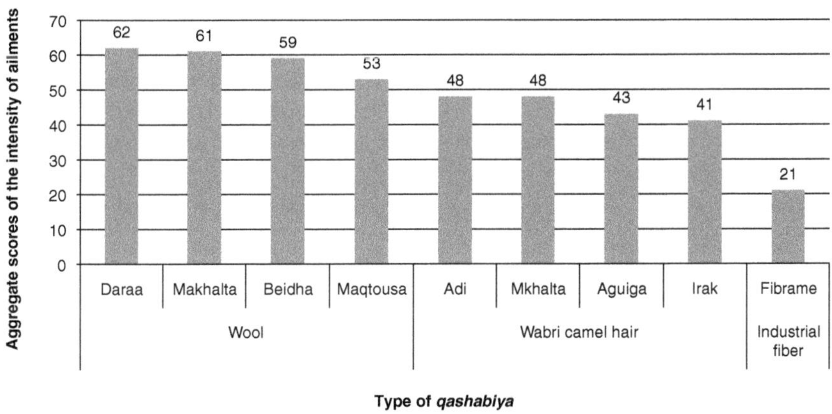

Figure 5.2 Aggregate scores of the intensity of ailments by type of *qashabiya*

in rural areas. Therefore, women are exposed to less cold because the room's floor remains warm due to the type of apartments and in some cases due to neighbors' heating systems. Women in town work indoors which cause more allergies. Light in urban areas is better.

Through force field analysis, we identified the driving forces/perpetuating for these occupational ailments and discomforts: the unlimited working hours, the poor lighting, the unsuitable position of women while processing camel hair and wool, the use of traditional equipment for carding and more particularly spinning, the unsuitable position in weaving and the use of traditional looms that are unsafe for the entire household especially the children who play around the loom. On the other side, the restraining/mitigating forces were the weaving of industrial fiber such as fibrame, the use of masks to avoid dust while cleaning hair and wool (rarely seen in the study area, more used for dying wool), the use of chairs for weaving such as in other more advanced countries (Iran and Turkey) in terms of textile and wool weaving, the use of safe and modern equipment, and more importantly awareness of and information, as well as capacity building, on safety and the understanding of ailments' origins. By analyzing the gap between the driving and the restraining forces, one could improve the health conditions of the women weavers.

Traditional treatments of ailments

With the help of the problem tree, we identified some traditional treatments used in the study area to ease women weavers' ailments. Although they had the capacity to relieve few of the most pressing ailments, they remain a means of temporary relief that require replacement by more adequate means of preventive measures and treatment (Table 5.5).

Preventive measures

There are some measures capable of relieving women from many ailments. These could be effected during sorting of the raw material – wool, camel hair, goat hair and so on – or delivered through some sort of technological ventilation improvements to draw the dust outside the workplace. Another preventative measure would be to introduce better ventilation of the rooms in which the wool-washing and dying processes take place. Women could be trained and provided with rubber gloves and waterproof aprons.

Adequate lighting is essential in the weaving rooms. One of the main, inexpensive improvements could be the use of a stool such as in Iran and Turkey while weaving, or to raise the lower roller of the loom. This would help in redressing the back problems and avoid squatting on the floor in an unhealthy and uncomfortable way. Such an ergonomic improvement will not only improve the health of the women weavers but, once adopted, will undoubtedly increase their efficiency and productivity.

Recommendations and suggestions for technology development

As a summary, technology potential includes:

1 **Use of a stool, and other equipment.** Women sit on the floor while weaving. In Iran women use a stool, and feel much more comfortable while weaving. Most traditional equipment such as mchat (comb), khullala (weaving tool) andmaghzal (spinner) are still in use in the study area. There were some attempts to improve the maghzal (spinning tool), but the modest electrical

Table 5.5 Traditional treatments used by women for occupational health ailments

Treatments \ *Ailments*		*Joint diseases*			*Allergies/inflammatory*					*Eye disorders*	*Stress/ headache goiter*
		Sprain	*Pain*	*Infections*	*Cough*	*neezing*	*Throat*	*Gums*	*Paronychia*		
Traditional treatment	Use of *Henné*	X	X	X							
	Bandage/ medicinal plants	X	X	X							
	Hot water articulation massage			X							
	Irregular attendance to work/temporary rest	X	X								X
	Use of vinegar (Khol) on head										X
	Use of orange blossom water										X
	Hot bath		X								
	Hot sand		X								
	Ghee + hot milk				X		X	X			
	Olive oil massage			X							
	Coffee powder on the wounds								X		
	Khol									X	
	Use of a scarf										
	Sleep on hard surface										
	Olive oil (khullala)					X					
Medical treatment	Anti-inflammatory										
	Prosthesis										
	Kneepads										
	Belt										
	Ankle pads										
	Cortisone										

* Paronychia (Cuticles pain)

spinning tool designed by local people was considered potentially successful for wool but not for camel hair that needs to be very thin. Women have some resistance to change, especially when it comes to trying one specific phase of wool and camel hair processing. They want to be assured about the result because of the high investment into previous phases of processing which they know delivers results, and they cannot afford to lose money. Entrepreneurial spirit is absent because women do not have any insurance for trying something new, and also because they are operating in isolation. The two horizontal wooden parts of the loom are attached on the one side to the upper level of the loom with two perpendicular wooden bars (manouchate) fixed to the wall and on the lower level by another wooden part (manetwa[10]) supported with a big rock or another heavy item to keep it fixed to the floor. If this stone is pushed inadvertently it sometimes slips and hits women in the face, causing retinal detachment or hits children playing close by. Khachab (wooden bars) cost 7000 DA, and many women cannot afford to buy them, they borrow them from others. A long metallic bar fixed along the wall, that is used to support the loom and as a hanger for clothes or to dry clothes during winter, now replaces the traditional manouchate in some areas.

2 **Protecting equipment.** Promote the use of knee and neck protection (knee pads, neck brace), belts, and masks when cleaning wool. A woman has lost her voice as a consequence of boiling wool or pouring boiled water on wool for better cleaning, because she used a product called 'Nasseh'[11] for cleaning wool. Its mixture with hot water generates a harmful steam that is inhaled by women. Alternatively sulphur was previously used to whiten the wool, but nowadays, it is not available in the market.[12] There needs to be a campaign to raise awareness in those women, through public health institutions and media. Medical doctors are ready to collaborate with researchers to formulate adequate solutions to prevent women weavers' ailments.

3 **Electric spinning tool**. There were some attempts to improve the maghzal (spinning tool), but the modest electrical spinning tool designed by local people was considered potentially successful for wool but not for camel hair that needs to be very thin. However, women have some resistance to change their traditional habits simply because they usually work only on one item, and its income is used to produce the next one. Therefore they are reluctant to try a new technology that may affect their product. Furthermore, they want to be sure about the successful result from the electrical spinning before trying it themselves. This could be arranged through traveling workshops to countries where it is in use, to demonstrate its efficacy.

4 **Washing machine for wool**. To avoid respiratory diseases due to inhaling wool washing residues during boiling, an individual or collective washing laundry could support women workers in rural and urban areas; there is a good case for designing special washing machines for wool, either for individuals or a collective.

5 **Modifying the finishing/polishing (hack)**. It is actually done with a piece of palm tree that affects women's fingers when polishing the *qashabiya* cloth; this causes substantial pain. Alternatively, the same piece from a palm tree could be improved with a handle to protect women's fingers.

6 **Vaccination.** Women could be vaccinated against allergies (tests, treatments etc…).

Involve the state and professional bodies in bringing about change

To ensure and protect the reproduction of *qashabiya*, the programs of professional training should consider including *qashabiya* in their courses.

The research has identified the cause–effect relationship between making *qashabiya* and the ailments that affect women. The Ministry of social affairs and labor, and the Ministry of health could follow up by making an inventory of the ailments and diseases of home-based women as professional diseases. This classification would allow them to claim benefits for curing their ailments as they would be covered by insurance and social security.

There were a few efforts made by the state to encourage and support rural women's associations to take the lead for the development of these activities and contribute to capacity building in this sector, in collaboration with a Spanish development project (Thieux, n.d.), but they remain insufficient.

The Ministry of crafts and trade should consider *qashabiya* production as a formal professional activity although it takes place at home, and allocate related benefits from development programs such as credits, information and safety.

Empowerment

Awareness of women weavers about their health issues (power from within or perceptual change – bottom up)

The majority of women interviewed had never manifested their concerns about health problems related to their home-based weaving activities, simply because they were raised within an environment where these health issues are part of normal life, especially because they perform many painful activities to secure livelihoods. From a bottom-up approach, their awareness needs to be increased through exposure to more choices and options, as well as increasing their confidence and widening aspirations. They will be able to increase their 'power within' that refers to self-awareness, self-esteem, identity and assertiveness (knowing how to be). Power within also refers to how individuals, through self-analysis and internal power, can influence their lives and make changes (Charlier *et al.* 2007). The women's confidence in the quality of work they perform is not questionable; however, their confidence in speaking out about concerns, aspirations and options is still weak mainly because they are confined to the household boundaries and narrow neighborhood that offers low potential for change.

Awareness of medical doctors about the relationship between the women's occupation and related ailments (power from within or perceptual change)

Medical doctors visited in the governorate do not make any relationship between the ailments of patients and their occupation (no cause–effect relationship). *Qashabiya* making is a home-based activity, therefore not accounted as an income-generating activity such as weaving within a factory, and seen as such by medical doctors. Most autochthon doctors understand the relationships between the activities and ailments, unlike others raised in different environments. A few doctors showed interest in working more with and be alerted by researchers through jointly developing appropriate questions to cover the different types of occupation in the informal sector.

The few autochthon doctors that originated from the study area were aware that the ailments of women weavers are related to their activities. Their awareness has beens generated from the fact that they were raised in similar environments where weaving was the main activity for their mothers and women relatives. However, they had never thought that health issues related to weaving were important enough to be raised at a national level, mainly because women were not officially employed in a public or private sector nor registered into any health insurance scheme.

Doctors need to be alerted by rural development researchers and development agencies. Most of those contacted have expressed their willingness to cooperate with researchers in two ways: 1) better understanding of the origins of women's ailments though the preparation of better questionnaires to patients investigating more deeply about the occupation of the ill women even when they declare 'not working and staying at home'. 2) Raise the issue at national level through collaborative initiatives. Such a strong stance could lead to an important research and development initiative.

Conclusion

Examination of the occupational issues of *qashabiya* production reinforces our hypotheses and results about the situation of women processors of raw material for weaving and weavers. It is clear from the research that it is very difficult for women in these communities to manage the health risks associated with the traditional fiber industry under complete absence of insurance cover and safety regulations. Women producers of *qashabiya* are regarded as home-based, non-productive individuals only because they are working at home. These women are contributing to the national economy and sustaining a symbolically powerful, traditional garment which is their national heritage, and their work also improves the livelihoods of many rural households. The situation looks even worse as such an important profession, preserved over generations, is not regarded through a legal lense simply because it is performed at home by powerless women. Considerations for an inclusive growth strategy are highly recommended.

Notes

1 International Center for Agricultural Research in the Dry Areas (ICARDA), Amman.
2 École Nationale Supérieure d'Agronomie (ENSA), Algiers.
3 Institut National de la Recherche Agronomique (INRA), Djelfa.
4 Aspects of power include: the 'power over' that involves a mutually exclusive relationship of domination or subordination, and is exerted over someone where it allows 'someone to be guided'. The 'power to' includes the ability to make decisions, have authority, and find solutions to problems, and can be creative and enabling. The notion

therefore refers to intellectual abilities (knowledge and know-how) as well as economic means, i.e. to the ability to access and control means of production and benefit (the notion of assets). The 'power with' refers to the social or political power which highlights the notion of common purpose or understanding, as well as the ability to get together to negotiate and defend a common goal (individual and collective rights, political ideas such as lobbying, etc.). Collectively, people feel they have power when they can get together and unite in search of a common objective, or when they share the same vision. The 'power within': this notion of power refers to self-awareness, self-esteem, identity and assertiveness (knowing how to be). It refers to how individuals, through self-analysis and internal power, can influence their lives and make changes (Charlier *et al.* 2007).

5 Like many other governorates in Algeria, it is very common in Djelfa that the sons share their parents' house for years before deciding to live separately. Often this is dictated by parents and also allows the husbands to work in the cities and abroad leaving their families under the protection of the elders. This reinforces the castration of women within the house in these conservative societies.

6 Hand carders come in a wide variety of sizes, from 2 by 2 inches to 4 by 8 inches. The small ones are called flick carders and are used just to flick the ends of a lock of hair, or to tease out some strands for spinning off. http://www.innovateus.net/innopedia/what-does-carding-wool-mean

7 Although this was not the focus of our research, discussions with key informants indicated that chemical residues of the nuclear center located in Reguane and Birine villages, Oussera disctrict in Djelfa governorate 100km from the research site may have had an impact in the area: two nuclear experiments were conducted during the French occupation of Algeria. Rangelands and camel hair in Reguane may have been affected by radioactive residues, transferring their effects on women weavers who process camel hair to produce *qashabiya*. They were mainly affected by goiter, which may also be attributed to lack of iodine in water.

8 Magatos is the name for a local product used for camels for scabies and chips. Nowadays, there is a veterinary product on the market called Sedacil that is also used by livestock producers and sometimes also called Magatos because used for the same purposes.

9 The name of an Algerian detergent used for bathroom cleaning.

10 Manatwa is a large wooden bar on which the woven part of the *qashabiya* is folded. It is dangerous for people working close-by because it is maintained by a heavy big rock, which if moved can cause disasters, int has in many cases caused retinal detachment when hitting the faces of women.

11 This product is a detergent. It is used for wool because sulphur is unavailable in the market since the 'black decade' of political turmoil in Algeria in the 1990s.

12 It was used by Muslim fundamentalists to process explosives during the 1990s and 1980s.

References

Agnihotram, R.V. (2005) 'An Overview of Occupational Health Research in India'. *Indian Journal of Occupational and Environmental Medicine* 9(1), 10–14.

Alary, V. A, Nefzaoui, A. and Ben Jemaa, M. (2007) 'Promoting the adoption of natural resource management technology in arid and semi-arid areas: Modeling the impact of spineless cactus in alley cropping in Central Tunisia'. In *Agricultural Systems* 94, 573–85. Elsevier, www.elsevier.com/locate/agsy.

Alkire, S., Meinzen-Dick, R., Peterman, A., Quisumbing, A., Seymour, G. and Vaz, A. (2012) 'The Women's Empowerment in Agriculture Index: Preliminary Draft released as Research in Progress'. Online at: www.ophi.org.uk/wp-content/uploads/OPHI-RP-35a.pdf?cda6c1.

Becker, E. (2012) 'Themes from Feed the Future Women's Empowerment in Agriculture Index: Report from Qualitative Case Studies in Bangladesh, Guatemala and Uganda'. Unpublished report submitted to the International Food Policy Research Institute.

Charlier, S., Caubergs, Malpas, N. and Mula Katiba, E. (2007) *The Women Empowerment Approach: A methodological guide*. Commission on Women and Development.

Chevalier, J. and Buckles, D.J. (2008) *SAS²: A Guide to Collaborative Inquiry and Social Engagement*. IDRC with Sage, Thousand Oaks, CA.

Houba, R., Doekes, G. and Heederik, D. (1998) 'Occupational respiratory allergy in bakery workers: A review of the literature'. *American Journal of Industrial Medicine* 34(6): 529–46.

Islam, S.S., Velilla, A.M., Doyle, E.J. and Ducatman, A.M. (2001) 'Gender differences in work-related injury/illness: Analysis of workers compensation claims'. *American Journal of Industrial Medicine* 39(1): 84–91.

Mahmud, S., SHAH, N.M. and Becker, S. (2012) 'Measurement of women's empowerment in rural Bangladesh', *World Development* 40(3): 610–19.

Mayoux, L. (2003) 'From marginalisation to empowerment: towards a new approach in small enterprise development'. Paper presented to SDC Employment and Income Division Workshop.

Messinga, K., and Stellmanb, J.M. (2006) 'Sex, gender and women's occupational health: The importance of considering mechanism'. *Environmental Research* 101(2): 149–62.

Motamedzadea, M., Choobinehb, A. Mououdic, M.A. and Arghamid, S. (2007) 'Ergonomic design of carpet weaving hand tools'. *International Journal of Industrial Ergonomics* 37(7): 581–7.

Thieux L. (n.d.) Rapport sur le Mouvement Associatif et la Femme dans le Milieu Rural en Algérie. Institut d'Études sur les Conflits et l'Action Humanitaire. (IECAH). Agence Catalane de Coopération au Développement, CERAI.

6 Access of rural women's cooperatives to markets in Lebanon

Barriers, enablers and options for action

Lina Abou-Habib[1], Omar Traboulsi[1], Aziza Khalidi[1] and Habbouba Aoun[2]

Introduction

Rural Women's Cooperatives (RCW) represent an emerging institutional pathway towards financial autonomy for women and a component of a marginalized agricultural sector. In Lebanon, a rural cooperative is defined as 'an autonomous association of ten persons with a capital that is not limited; it is non-profit, and its aim is to improve the socio economic conditions of its members through joining their efforts' according to the general principles of cooperation: Legislative Decree 17199 dated 18 August 1964.[3] Originally, rural cooperatives were organized by a government decree in 1941 and overseen by the Directorate of Cooperatives under the Ministry of Agriculture. Today, in a less-than-optimal arrangement, oversight is distributed among several ministries including the Ministry of Industry, Ministry of Social Affairs and the Ministry of Agriculture.

Despite seemingly favorable conditions, such as availability of water, temperate climate, and access to arable land, the agricultural sector in Lebanon is relatively underdeveloped compared to neighboring countries. The current status is a product of the historically low government priority attributed to the sector, and the lack of coherent development strategies to promote growth and address structural deficiencies.

Although the Lebanese food industry has consistently increased its volume of exports over the years, there are still many internal issues impeding further progress. For instance, trade liberalization is expected to open many new markets for Lebanese products nevertheless it may also increase competition from imported goods as the Lebanese market opens to international markets. Particularly important is the need to increase food quality and safety of products to meet international standards. Several agreements on Sanitary and Phytosanitary (SPS) and Technical Barriers to Trade (TBT) are being considered.

In addition to policy and institutional challenges, several socio-economic realities impede women's attainment of financial autonomy: according to 2010 estimations women only make up about 27.7 per cent of the workforce.[4] In addition, only 7.2 per cent of the total labor force in Lebanon works in the agricultural sector (4.6 per cent of the females and 10 per cent of the males).[5] According to a recent FAO publication, women constitute 9 per cent of farmers.[6] Such a low proportion of participation is attributable to women not returning to the work-force after marriage and child-bearing due to prioritization of household duties. In addition, there is a general reluctance from employers to hire women. In rural areas in particular, employers prefer to hire migrant and informal workers. Informal, temporary workers reduce costs to the business because they go unreported to the government, thus allowing the employer to evade taxes on their wages.

Study objectives

The goal of this study was to generate information that would improve access of RWCs to the market. The operational objective was to identify:

- barriers that impede access of Rural Women Cooperatives to markets
- factors that enable access of Rural Women Cooperatives to markets
- options for prospective interventions to improve positioning in the markets; taking into consideration lessons learned from experiences.

Study area

The study area covered different regions of the country, but mainly in: Bekaa (Zahleh, Rashaya), South Lebanon (Tyre), and Nabatiyeh (Nabatiyeh). Communities included in the research are generally classified as rural agricultural areas with poor access to basic services including education and health services. Socio-demographic data available about the different Lebanese regions showed that poverty is highly reported in certain areas of Bekaa, Nabatiyeh, and South Lebanon. Women with disabilities are among the poorest in the country. Poverty correlated with a number of characteristics including family size (the poorer the family, the larger the family size), educational attainment (the poorer the family, the lower the educational attainment), as well as age and sex of the head of the households (deprivation rates were higher among female-headed households). All three areas included in the research were agricultural, and it was found that poverty was more prevalent among agricultural workers, the majority of whom were illiterate.[7]

In Lebanon, there still exist large regional disparities in access to education, despite national efforts aimed at expanding access to public education, mainly in poor rural areas.[8] Available statistics revealed that illiteracy was most prevalent in older women (40+ years), particularly from rural areas,[9] despite Lebanon's efforts towards achieving gender equality in educational attainment.[10] In general, there appeared to be no gender disparities in access to primary education. However, girls and women in rural areas, who are among the marginalized and vulnerable population groups (including agricultural workers), continued to comprise the majority of out-of-school children and adult illiterates, respectively.[11]

Usually, economic participation in Lebanon is known to be highest in Beirut (51 per cent of Mohafaza residents), followed by Mount Lebanon (47 per cent), and then North Lebanon (40 per cent).[12] Female economic participation was reported to follow a similar trend with highest participation in Beirut (36 per cent), followed by Mount Lebanon (24 per cent), then by Nabatiyeh (19 per cent).[13] More thorough analysis of female economic activity showed that 69 per cent of employed females were in Beirut and Mount Lebanon, 11 per cent were in North Lebanon, 7 per cent were in Bekaa, and 13 per cent were in South Lebanon and Nabatiyeh.[14] Female unemployment was highest in South Lebanon and Nabatiyeh, followed by Mount Lebanon, Beirut, North Lebanon and Bekaa.

Access to health services across the country is disproportionate due to geographic, financial and other barriers. The regions of Bekaa (Baalbeck and Hermel), North Lebanon (Akkar, Minieh and Donniyeh) and parts of South Lebanon (mainly Nabatiyeh) are among the most deprived in Lebanon. In these regions, 60–70 per cent of the residents do not benefit from any health insurance programs.[15]

In context, RWCs participating in the study came from communities classified as poor rural areas with limited access to resources, including education and health services. Women's economic participation in these areas was low, as it was a low priority Thus, the establishment of RWCs came as a much-needed benefactor to the whole family unit and not only the women concerned.

Methods

Qualitative methods of participatory action research were used. Analytical tools included a timeline, SWOT analysis, a causal dynamics tool, a social domain approach and the Socratic wheel. The main topics of discussion were framed in five key questions detailed in Table 6.1.

Data collection took the form of a series of two-day workshops. These workshops were conducted across the four study areas with members of the cooperatives. Workshops grouped representatives from 25 rural women's cooperatives. The research workshops took place between April and May 2011 (see Table 6.2). Data were recorded, transcribed, compiled and analyzed.

Table 6.1 Overview of general topics addressed by analytical tool used

Question	*General topics addressed*	*Data collection tools*
1	What are the main constraints faced by the RWCs in accessing internal and external markets?	Timeline tool
2	What are the strengths/assets and weaknesses of RWCs?	SWOT analysis
3	How do institutional policies affect RWCs in accessing markets and other services? (Refers to external context)	Causal dynamics tool
4	What are the present processing, market and management skills of the RWCs, and how do they affect their ability to access the markets?	Social domain approach
5	What are the interventions (skills, information, others) which can facilitate RWCs in entering new domestic and international markets?	Socratic wheel

Table 6.2 Field work schedule by location, number of participating RWCs and number of individual women participants

Date	*Location*	*Number of RWCs participating*	*Number of individual women participants*
8–9 April 2011	Rashaya	5	14
13–14 April 2011	Zahleh	8	21
29–30 April 2011	Nabatiyeh	7	19
6–7 May 2011	Tyre	5	10
	Total	25	64

Results

Barriers impeding organizational growth

Membership attrition

Membership attrition occurred in most RTCs in the study. Early dropouts from membership were based on unrealistic expectations 'People thought they would produce (make a profit) immediately' as in the case of Bint Jbeil RWC. Lack of liquidity to spend on initiation costs such as purchasing fuel or seeds contributed to dropping out. According to Bint Jbeil RWC, when women realized they had to pay from their own pocket and they expected to make money they dropped out. It is worth noting that in Bint Jbeil some of the dropouts who took the training worked on their own and that was considered in general a positive step by the RWC but it did not prompt examining what could be done to entice women to return to the RWC arrangement. A painful dropout experience was with the Kweikh RWC in Bekaa where the president quit because of personal reasons. She was deemed most competent. This was considered to be a setback but the rest of the team managed to continue. The Kweih RWC has an excessive number of unsold products which it recognises as a marketing problem.

Aiha RWC is another example of attrition in membership. It started with 75 members but when work commenced and the women realized that it was not an employment with monthly pay, they started dropping out. However Ein el Loz RWC in Ain Ata, Rashaya area is a more positive example with relatively low attrition from 26 to 23 with a core of 7 most productive members.

Armed conflict

In the south, the 2006 war and resulting displacement and destruction of premises of coops delayed the process of maturation of the RWCs. However, all cases in the south of Lebanon reported resilience: a set back after the initial shock was followed by a rebound with the support of civil society and local governance structures.

Lack of clarity during initiation phase

The Jdita RWC is an example of passive involvement of women in the process of initiation. What sustained this small group of women was a sense of traditional pride. The women who are running the RWC were not the decision-makers during initiation. The end result is that the RWC is barely running.

Lack of market insights

The experience of Bint Jbeil RWC shows that there is an implicit doubt on the part of women of the capacity of new crops to replace the traditional tobacco agriculture in profitability. In fact tobacco agriculture appears to be the most reliable option that is available when all else fails. The negative experience of Debel RWC attests to that matter. Lack of good market assessment before starting a product line led to negative experiences and to shifting to new product lines. For example, the Khirbet Kanafar experience shows a choice of product that was not based on solid knowledge of the market. Women started with crafts and then they discovered that they had strong competition from Syrian and Chinese crafts, so they decided to shift to the traditional food industry 'mouneh'.

Lack of follow-up by support agencies following initial training

As indicated by Bint Jbeil RWC, the need for 'supervision' was considered vital, and the lack of such resulted in financial losses.

Loss of premises

The Debel cooperative in the south had a strong initiation and motivation; and was extensively supported, thus becoming very productive. However, the cooperative did not own the physical facility where their center was located. This RWC survived the aftermath of the 2006 war. However, the owner of the building decided to regain the premises for purposes of construction. The work stopped because the machinery had to be dismantled and now they are searching for a storage facility. The morale is low. They have returned to tobacco agriculture. RWC members are greatly disenchanted because they enjoyed working in an RWC; in addition, they have a broad range of equipment as well as know-how and outreach capacity to market.

Incomplete supply of equipment by donors

The Bint Jbeil RWC[16] experience was of an increasing supply of perishable products, seedlings (fruit trees, decorative plants, selected legumes) kept under tents, exposed to the weather conditions in the area. According to the RWC respondent, women were only able to cultivate one season out of two per year in the greenhouse because of lack of heaters to sustain the crop during the cold season. The women had to reduce their expectation of making profit over two seasons as they could only produce enough yield in the summer.

Degradation of products and materials

A growing inventory of finished products, or sometimes accumulation of extra soil for excessive periods leads to a loss of potency, hence incurring financial losses as experienced by Bint Jbeil RWC.

Disempowering perceptions of women

Disempowering perceptions of women's capabilities are reflected in a subtle manner in almost all statements by RWCs. The most prominent was made by a founder of Al Ain RWC.

> We faced a challenge, the word spread in the village that this cooperative is solely run by women; they do not have a man with them. They are overestimating themselves (and challenging men)!! Without having men around them, leave them for a week or so and they will surely close. In a month's time they will disintegrate. Women are known to quarrel among themselves!!

Another gender-based disempowering factor was voiced by the president of Aiha RWC in Rashayya area (southern most area of Bekaa). Women's movements are restricted. Women are not allowed to go out of their home in the evenings and go to exhibits to market their product. She is the only one who is able to move and that inflicts a burden on her.

Factors enabling access to market

In addition to overcoming barriers of access to market, study findings revealed several factors that enable access to markets. In the presence of political instability and armed conflicts, which can be considered as factors that are beyond the control of the cooperatives, internal organizational factors emerge as the prime enabling factors. Such organizational factors consist of resources available – financial, human, material, informational as well as organizational – policies and procedures and a system of governance. Also important is the set of values that prevails within the cooperatives.

Women's leadership

The Harees RWC (South) and Al Ain (Bekaa) experiences are two cases illustrating a type of leadership that fosters cohesiveness based on sensitivity to the situation of rural women, being women as well as being rural. A similar situation was voiced by Al Ain RWC. Also such leadership exercised a set of business decisions that showed high self-esteem coupled with business acumen and indicated gender empowerment. The importance of having a woman running the business was voiced by Irsal RWC during the timeline account. However, this was intertwined with poor performance that brought financial losses, so it is not clear whether the most important issue was for women to be in decision-making positions or in fact to seek a good performance.

Lack of women's leadership had a derailing effect on the growth of the RWCs. In the case of Rashaya cooperative the husband of the previous director was the one involved in launching the project in his capacity as the head of the municipality along with the funding agency. His wife was appointed as director with a monthly salary of 300 US dollars according to the statement of the current president. This situation created a problem of 'hegemony' for five years from 2003–2008, as the current president stated. However, women organized an 'uprising' where they regained their power through democratic processes where three out of five members of the board cancelled their proxies to the head of the municipality, so he lost his majority. Currently Rashaya is a regarded by the president as successful. They have access to external markets.

Sustained motivation of an individual or group

A core individual motivator was the glue that sustained continuity of work for Al Ain RWC in Bekaa. One woman through her persistence kept the organization going as membership participation fluctuated. A similar situation was evident from the experience of Ein Arish RWC in Bekaa where the founder individually financed the cooperative by paying the seed money and registered the cooperative in the name of 11 women. 'Morning Star' RWC in Muhaidtheh village in Rashayya area is a story of sustained motivation during the start-up phase. Such motivation was manifested by investing time. One member even offered a modest space for use as a center during the start-up phase for minimal fees. That RWC was able to export its products and maintain a broad network base.

Support of national and local governance structures

Support from local governance structures is a factor that permeates the experiences of most RWCs. Successful RWCs such as Deir Kanoun and Harees reported tangible support from the municipality. The demise of the Debel RWC raises questions regarding the responsibility of local governance structures to sustain RWCs. Although discussions did not elaborate on the topic, the silence itself provokes questions on its contribution to this demise.

Prime focus on integrated organizational development vs. specific activities during initiation phase

The process of initiation of cooperatives may shed light on the sustainability of the initial motivation. At the Shehabieh RWC, the focus was on establishing the organization, women took the initiative and contacted the YMCA to establish an RWC, while in Bint Jbeil the motivation was focused on switching from tobacco to alternative crops and Sadr foundation was contacted for training to meet that objective.

Harees RWC experience sheds light on the matter of 'volunteering' in the initial stages, a core of five women did not take any remuneration during the start-up stage, while they noted that other women wanted to be remunerated on each simple activity they did. The point is that the mix of the team at initiation who are aware about the importance of investment of time contributes to reducing start-up costs.

Growth within a gendered framework

Deir Kanoun and Harees RWC are examples of this but with different configurations. Both are resilient and growing RWCs.

Deir Kanoun RWC experience is a case in point – an organization with an increasing trend of membership (12, 18, and 23 members at the time of the study) – where women indicated that although their initiation was not focused organization building, their active interaction with the support organization prompted initiation. Furthermore, in their accounts, women elaborated on the specificities of the situation of women and the interplay between absence of autonomous income and their intense attachment to family especially following the war.

These women were on the lookout to diversify without committing the organizational structure. The Zaatar growing experience is a case in point where the RWC women allied themselves with other women who had land but did not have liquidity and knowhow. They provided them with the ingredients. They worked together without incorporating them as members to avoid overexpansion.

The Harees RWC women focused on supporting young women who married early and who also dropped out from school as a factor in initiation, which shows the presence of an underlying basic 'genderized' frame. This was complemented by a leadership style that focused on 'protection' of the younger, inexperienced members during the trips to exhibits. Undoubtedly this had a positive implication on the image of working women in a traditional rural environment. This frame is coupled with indications of market acumen especially in marketing. The women worked competitively by focusing on quality. They managed to secure a local outlet in Harees and can afford an employee on a monthly salary. Harees RWC management was aware of the advantage of having a local outlet to attract local buyers and compete with home-based catering. The use of a RWC business card as a marketing tool is also another indication of success in seeking opportunities. They have an independent business locally and in the capital, in addition to the traditional venue of the seasonal joint exhibits.

Another strategy of opportunity generation is product related, drawing from the tradition of 'travel bread' made of sesame, flour and oil, which has long durability compared to other forms of bread. Furthermore it has an added value of being a unique, traditional product which has its specific market and it is economically feasible. The women took the decision to produce that product and focus on 'traditional food' as a market niche.

Specialization

The experience of Ain Arish cooperative in Aita Fakhar, Bekaa area of focusing on one product: goat yogurt cream, 'labneh', which was associated by the respondent – the founder of the RTC – with greater marketability, had led to the success of this RWC. She indicated that 11 members now have financial returns, have become more integrated in society, and the RTC has relations with several NGOs and has participated in many exhibits.

Options for action

Options for action are based on the reported experiences of RWCs.

Conduct regional market studies

Observations of surrounding markets were exercised by several RWCs – some RWCs such as Bint Jbeil RWC stated that clearly in the workshops while for others, such as Zahleh and Harees RWC, it was implicit.

Establish linkages with broader markets

Zahleh RWC emphasized the importance of having a regional outlet. Firzol RWC in Bekaa, a productive and experienced RWC, voiced the need to link with companies for export given their capacity to produce and their confidence in their capability to adhere to international standards. This strategy is consistent with the statement of Irsal RWC which underscored the importance of seeking markets outside the immediate locale and the region to achieve economic feasibility. It is also consistent with the experience of RWCs in the south where they sought linkage with the broader Lebanese market.

Facilitate access to media

As was suggested by Zahleh RWC, access to media is necessary both for marketing purposes and increasing public awareness. It serves communities as it means that national and international public institutions will consider RWC when providing development related support.

Design awareness-raising strategies targeting communities

Community awareness-raising about cooperatives by outside specialists was voiced as a recommendation by Khirbet Kanafar RWC. Kawkaba RWC expressed the need for participants in rural cooperatives to be committed to the idea of a cooperative: 'You need to believe in the cooperative way of conducting business before committing yourself to this work. You have to believe that you are entering to build yourself, to grow' Kawkaba RWC member.

Conclusion

The study shows through first-hand accounts of successful experiences the feasibility of establishing rural cooperatives run by women. Knowledge of and commitment to principles of cooperatives were common characteristics of RWCs who are currently doing well, that is those that have access to market. The notion of being not-for-profit as well as requiring a start-up period of low return of investment on time and energy and an understanding of limitations to movement of women need to be factored in in future development initiatives. Monitoring and support during the start-up period need to be factored into the plans of action at the initiation of future cooperatives. Special avenues of support should increase awareness on the principle of equality as articulated by the Convention on Elimination of All forms of Discrimination against Women (CEDAW) along with management and marketing capacity building and technical monitoring regarding quality of product.

Annex 1: Data Collection Tools

Question 1: What are the main constraints faced by the RWCs in accessing internal and external markets?

A timeline[17] was drawn and a historical review and analysis of main events and their significance was conducted. Participants were asked to trace back the history of establishing the cooperatives and to answer questions including the following:

- When were the rural women cooperatives established?
- When – since their establishment – were they first funded?
- What were the main landmarks during the lives of the rural women cooperatives?
- Why are these landmarks important?
- What were the events that occurred during those times?
- How did these events affect the performance of the RWCs?
- How did these events impact their ability to access the markets?

Participants were asked to create a 'before-and-after' table on large chart with five columns. Column 1 described 'areas of change', column 2 described 'what used to be'; column 3 described 'the present situation', column 4 described for each area of change, how important these changes are, and column 5 described the causes or reasons behind each change. Finally participants were asked to describe the overall difference between the past situation and the present. These procedures were organized in the small working groups (four groups of four to five participants), followed by presentations and discussions.

Question 2: What are the strengths/assets and weaknesses of RWCs?

This question was dealt with by SWOT[18] (strengths, weaknesses, opportunities and threats) analysis. These aspects were asked on both external and internal factors perceived by the research participants as having had an impact in shaping their experiences and their progress.

Question 3: How do institutional policies affect RWCs in accessing markets and other services? (Refers to external context)

The Causal dynamics tool[19] which is used to asses causes, was applied here. Whilst building on findings from the review and the external context, the inquiry focused on the following questions:

- How does the external context affect and shape the performance of RWCs and their ability to access markets?
- How does the economic context and institutional policy affect RWCs' abilities to access other services?
- To what extent do policies affect your access to market?
- At what level do the policies produce an obstacle for your accessing services?
- What are the policies that affect you?

This exercise involved defining key problems, listing the factors involved in a key problem, and designing the rating matrix and findings and conclusions. It took the form of brainstorming, small working groups, presentations and discussion.

Question 4: What are the present processing, market and management skills of the RWCs, and how do they affect their ability to access the markets?

The Social domain[20] approach was followed to address this research question. This technique is used to compare actors. The questions were:

- What is the size of the membership and number of workers in the RWCs?
- What is the profile, educational and social background of women members and workers of RWCs?
- What are the organizational management, production and marketing skills and know-how of RWCs?
- What are the gaps in each of the above that prevent RWCs from improving their access to internal and external markets?
- What are the advantages that can help improving RWCs' access to internal and external markets?

More detailed questions of the capacity of the RWCs included: how many of the members know how to use computers (beginners, intermediate, advanced); how many have marketing knowledge (beginners, intermediate, advanced); how many have knowledge of accountancy (beginners, intermediate, advanced); how many of them participate in sales and exposition; how many know how to use the food processing machineries available; how is the management organized; how many food processing machineries they have; whether they have a computer or not.

In the Social domain methods, the researcher uses terms and characteristics that participants choose and negotiate to examine how RWCs view themselves and others and similarities and differences between them. At the end of the exercise, participants discussed the main differences observed among groups and planned strategies that drew on different but complementary profiles. Brainstorming and small working groups were used in implementing the tools.

Question 5: What are the interventions (skills, information, others) which can facilitate RWCs in entering into new domestic and international markets?

In this case, the Socratic heel[21] approach was followed. This approach is usually used for ranking. The main question was: What is the impact of the various skills and know-how, on the ability of different RWCs to access the market?

Participants were asked to rate these impacts in order of importance. What are the most effective interventions to address the difficulties that RWCs confront in accessing markets? Participants were asked to rate these options in order of efficiency and feasibility. Brainstorming and small working groups were used.

Annex 2: Profile of the Rural Women Cooperatives that participated in the study

Zahleh

Name	*Date of establishment*	*Members*	*Main current production*	*Strengths*	*Weaknesses*	*Support received over time*
The traditional cooperative association of Kherbet Kanafar	1992	26	Dairy products, pickled vegetables	Well established cooperative has a good client base, mature membership	Difficulties in securing ransportation of products to Beirut and major cities, competition from commercial outfits (despite their lower quality products)	Various forms of training and capacity building, raw material and equipment
Rural Women Cooperative of Ferzol	1997	35	Dairy products, chick peas and cracked wheat, dried vine leaves	Secured HACCP standards, high quality production, established channels for local marketing	Difficulties in finding new markets, difficulties in passing on traditional skills and know-how to a younger generation (lack of interest)	Training in production, management, communication, accounting and marketing; raw material and equipment
Arsal Rural Food Production Cooperative	1999	27	Dairy products, various pickles, dried herbs, dried fruits, jams	Good relations with local community and with local and international organizations, starting to build their own premises, were able to have a contract to supply school meals for local schools	High membership turnover rate	Training, financial support, raw material, equipment, local support
Cooperative of young Rural Women in Al Ayn	2000	18	Various pickles, dried herbs, dried almonds	Potentially prepared to acquire the HACCP specifications, known for a limited variety of specialized products, equipment is available to be used by the local community	High turnover of membership because of low income	Several training workshops in production, equipment, raw material

Zahleh (cont)

Name	*Date of establishment*	*Members*	*Main current production*	*Strengths*	*Weaknesses*	*Support received over time*
Ain el Wardeh cooperative in Jdita	2007	7	Various herbs and pickles, jams, dried fruits, dairy products	Solid production know-how	Interference of political factions, dissent and turnover of members, no stable premises	Several training workshops in production, equipment, raw material
Al Sindian cooperative of Koueikh	2007	20	Traditional catering, hostel for eco-tourism, traditional hot meals, pickles and jams	Its specialty in eco-tourism.	Turnover and political interferences' several training workshops in production, equipment, raw material, belated marketing, activities leading to decay of products and financial losses	Several training workshops in production, equipment, raw material
Al Aahd rural women cooperative	2011	13	Dried fruits, sweets, jams and pickles	High quality products, good rapport amongst members, regular access to annual exhibitions which provide opportunities for significant sales	Limited marketing opportunities and no access to external markets	Several training workshops in production, equipment, raw material, capital support
Ein Arish Cooperative in Aita Fakhar	2003	11	Specialty: oat labneh (cream of yogurt), other products include grape leaves, pickles	Focus on a specialty product while maintaining production of a range of products on a smaller scale. Capable leadership. Core motivated members 7 of 11 are active. The product is known locally and in the capital.	No information is available on contribution of all members to financing, only the founder reported paying the dues of the 11 members	Several training workshops in production, Equipment, Raw material, Capital support, Use of avenues to reach local and national market such as basic word of mouth, networking with NGOs.

Tyre – 1

Name	*Date of set up*	*Members*	*Main current production*	*Strengths identified*	*Weaknesses identified*	*Support received over time*
Agricultural Cooperative for Developing Rural Women's Capacity in Bint Jbeil	Officially set up in 2009 but existing informally since 2007	40	Seedlings (fruit trees, decorative plants, selected legumes)	Strong collective and collegial work amongst women, demonstrable improvement in women's skills and ability to generate income from their work in the RWC, solidarity amongst women members and workers	Inconsistent supply of raw material, irregular marketing, inconsistent capital, water supply	Production training, seeds, tents
Shehabyeh Cooperative	2006	14	Specializes in 'victory biscuits, local catering, jams and preserves, dried herbs and leaves	Consistent training and capacity building, specialty products	The onset of the Israeli 2006 war right after the set up of the cooperative, irregular marketing access, major loss incurred when the cooperative produced a significant amount of products to a well known NGO marketing outlet which ended up not paying	Access to a major annual exhibition which provides a significant opportunity for sales
The Deir Kanoun cooperative for agricultural production (Ras el Ayn)	2005	23	Also specializes in 'victory biscuits'. The only one to produce the savory whole-wheat and multi-grain wafers, orange blossom, lavender and rose water, traditional finger food for catering	Consistent training and capacity building, evolving and diversified production, specialty products	Galloping costs of raw material, no access to external markets	Training, raw material, equipment

Tyre – 2

Name	*Date of set up*	*Members*	*Main current production*	*Strengths identified*	*Weaknesses identified*	*Support received over time*
Cooperative for Agricultural production in Debel	2005	14	Whole wheat, various kinds of pickles	Support received to re-establish operations after the damages incurred during the 2006 Israeli war, gradual diversification of production, participation in various exhibitions with reasonable sales	Galloping costs of raw material, insecurity related to their premises (may have to leave it soon)	Equipments, training, support from the local municipality.
Harees Rural Food production Cooperative	2005	22	Dried herbs, legumes, wheat and chickpeas, flower essences, dried spices	Support from the local municipality especially through providing a free of charge centre, training and equipment	Inconsistent marketing	Access to local markets and exhibition, training and capacity building, raw material

Nabatiyeh

Name	*Date of set up*	*Members*	*Main current production*	*Strengths identified*	*Weaknesses identified*	*Support received over time*
Arab Salim cooperative	2005	27	Jams, pickles, essences, specialty products: dried and candied wild oranges, catering	Products of high quality, women members very enthusiastic and motivated	No means of transportation, lack of specialty products, support not regular.	Training and capacity building, raw material and equipment
Kfarsir cooperative	2005	12	Specialty products: many kinds of vinegar and pomegranate molasses, jams	Internal skills and knowledge, good production capacity, Support of surrounding environment	Limited appreciation of healthy, natural and traditional products, intermittent market access	Training and capacity building, raw material and equipment

Rashayya

Name	*Date of set up*	*Members*	*Main current production*	*Strengths identified*	*Weaknesses identified*	*Support received over time*
Ain el Loz cooperative in Ayn Atta	2002	23	Pine kernels, dried chickpeas, dried green leaves, cookies	High quality products, well equipped, women work well together and play an important public role	Poor location very far from Beirut and other major cities, no regular markets, galloping costs.	Production training, accounting, ICT training, equipments, raw material
Kawkaba cooperative for women	2004	12	Pickles and jams, dried legumes and fruits, dairy products	Premises offered by the municipality for free, active and committed women membership, computerized systems.	No transportation means, no member with specific marketing skills	Several intensive, training and capacity building, raw material and equipment, access to markets, ICT training
Wadi Teem cooperative in Rashayya	2004	15	Herbal essences, sweets and cookies, dried fruits, spices.	Excellent relations with the local community and other local actors, a local client base, HACCP qualifications, known products with regular customers, car for transportation	Remote location of the cooperatives and prohibitive cost of fuel, no stable premises, no access to external markets	Several intensive training and capacity building, raw material and equipment, access to markets, ICT training

Rashayya – 2

Name	*Date of set up*	*Members*	*Main current production*	*Strengths identified*	*Weaknesses identified*	*Support received over time*
Ayha cooperative for traditional vissale products	2003	14	Jams and pickles, special essences, dried dairy products, dried legumes.	Permanent premises, very well trained and highly skilled members, high quality production.	No young members, no transportation vehicle, no marketing expertise, competition with commercial products	Several intensive training and capacity building, raw material and equipments, access to markets, ICT training
Nejmet el Sobh cooperative in Rashayya	2003	20	Biscuits, honey, dried and roasted wheat (a specialty product), dried fruits.	Well known by now and with a good client base, excellent working relations amongst members, a trusted leadership, computerised systems, all the members closely involved in production	Additional and new markets needed to enable the cooperative to increase production	Several intensive training and capacity building, raw material and equipments, access to markets, ICT training, support in expanding and refurbishing their premises

Notes

1 The Collective for Research and Training on Development – Action (CRTD.A).
2 University of Balamand.
3 Lebanese Ministry of Agriculture, Department of Cooperatives (2012) 'Al Daleel Al Irshadi lil Amal Al Taaouni Fi Lubnan' *Guide for Cooperative Work in Lebanon* (Arabic).
4 Estimations published by Yacoub (2008) Lebanon's Statistics on Informal Economy and Employment quoted by CRTD.A (2012) *Women and Economic Power in Lebanon: The Legal Framework and Challenges to Women's Economic Empowerment*, Figure 5, p.30.
5 CRTD.A (2012) *Women and Economic Power in Lebanon: The Legal Framework and Challenges to Women's Economic Empowerment*, p. 39 – Figures are from UNDP 2007 Household Living Conditions.
6 Food and Agricultural Organization of the United Nations – FAO (2012) *Lebanon Country Programming Framework 2012–2015*, page 5
7 Millennium Development Goals, Lebanon Report 2008, page 12.
8 Millennium Development Goals, Lebanon Report 2008, page 21.
9 Millennium Development Goals, Lebanon Report 2008, page 22.
10 Millennium Development Goals, Lebanon Report 2008, page 23.
11 Millennium Development Goals, Lebanon Report 2008, page 30.
12 'Living conditions of households survey of 2004', Ministry of Social Affairs, Lebanon
13 Ibid.
14 Millennium Development Goals, Lebanon Report 2008, page 30.
15 Millennium Development Goals, Lebanon Report 2008, page 39.
16 'Al Jamiyya al taawounia al zirariya litanmiat qodorat al mara'a fi qadaa bint Jbeil' – Agricultural Cooperative Association for Development of Capacities of Rural Women in Qada Bint Jbeil.
17 See *SAS² Guide*, p. 124.
18 See IFAD Gender Strengthening Program for Eastern and South Africa Division (2002) Toolkit for Practitioners – Gender and Poverty Targetting in Market Linkage Operations, p. 51.
19 See *SAS² Guide*, p. 140.
20 See *SAS² Guide*, pp. 86, 193–210.
21 Jacques M. Chevalier, Daniel J. Buckles (2008) *SAS²: A Guide to Collaborative Inquiry and Social Engagement*, International Development Research Center, A Sage Publication, p. 297.

Reference

Chevalier, J. and Buckles, D.J. (2008) *SAS²: A Guide to Collaborative Inquiry and Social Engagement*. IDRC, Sage: Thousand Oaks, CA.

7 Factors of market access success for the traditional food industry in Lebanon

Some themes pertinent to rural women's cooperatives

Lina Abou-Habib[1], *Omar Traboulsi*[1], *Aziza Khalidi*[1] *and Habbouba Aoun*[2]

Introduction

Mainstreaming gender into research on organizational performance is an emerging area of research. Bird and Sapp (2004) identified a gender gap in small business success favoring men in urban and rural comparisons. In an earlier study Kalleberg and Leicht (1991) analyzed indicators of survival and success of small businesses including food and beverage industries in the US and noted that the processes are similar for businesses headed by women and men, however, they detected preliminary indications that 'Different market approaches may pay off differentially for women and men'.[3] The authors considered the flexibility imparted by a generalist orientation to product portfolio and an emphasis on quality as factors that are 'more salient to the survival chances of businesses headed by women'.[4]

Since markets have been increasingly globalized due to free trade and to the enactment of several regional and multilateral trade agreements, food industries are facing competition more than ever. Despite that context, several Lebanese companies have been prepared to adapt and taken advantage of the open market despite the food trade imbalances that free trade has imparted. In addition rural women's cooperatives are seeking market niches in such a competitive environment.

This chapter attempts to shed light on the role that gender issues play in the success of such organizations as measured by access to local and external markets. Building on findings from five case studies conducted with five companies that are similar in size and product line to RWCs, this chapter explores the patterns of utilization of women's capacities in leadership and operation as they interplay with organizational and contextual factors towards the success of small businesses in accessing markets.

The following sections cover the methodology used to select and access the five companies, general profiles of the five companies and finally the themes derived that are relevant to Rural Women Cooperatives.

Selection and access to participating establishments

Several industries were chosen for the case studies. These industries were selected from a list taken from the Lebanese Syndicate of Food Industries.[5] The selection was based on the following four criteria:

- First, all companies should produce traditional products similar to those produced in the RWCs. Among others, these products ranged from pickles, jams and preserves, syrups, distillates to any other labor intensive fruits or vegetables, packed products such as canned grape leaves, packed dried chick peas, etc.
- Second, companies were chosen based on their geographical locations. Some of them were specifically chosen in rural areas while others in urban or suburban locations. These varieties provide a wider perspective on the role of women and the challenges or opportunities of market access in each location.
- Third, these companies should be small or medium-sized since in general (SMEs) face different challenges compared to larger organizations (UNIDO 2009).[6] SMEs reflect some common challenges and opportunities that could be also faced by RWCs.
- Fourth, the candidate companies should be successful in the market. This study relies on one or a combination of factors as measures of success – the company's overall brand reputation, the existence of products on supermarket shelves, the number of years and sales growth after establishment, and information gathered from the syndicates or associations pertinent to these industries.

Seven companies were selected for interviews. The majority (five out of the seven) accepted to be interviewed. Reasons reported for not participating were time constraints and lack of interest. All interviews were on-site. The data collection instrument was a set of semi-structured questions covering five topics: marketing strategies, product development, quality control, relationship with the RWCs, and gender. The subjects interviewed were either general managers or one of the owners or shareholders. Each interviewee was visited either once or twice depending on availability and research needs. The duration of each interview lasted between 60 and 100 minutes. Four of the interviews were recorded with the permission of the interviewees. Transcripts were later prepared for each recorded interview.

Limitations included interviewing only one person from the establishment. This person may have had a selective perception of the company's success and women's role in the food industry. Furthermore, the interviewee may have recall bias. Some might also use the case studies as a means to advertise their products hence reducing emphasis on hurdles faced or other negative factors. All of these factors could possibly reduce the authenticity of answers and the reliability of results.

However, questions posed in these case studies as well as the analysis were specifically designed to minimize the above-mentioned biases. Questions did not target issues related to accounting and finances, wages in real values, and names of dealers and suppliers where considered confidential. Also, most of the questions did not

Table 7.1 Profile of the five establishments included in the study

	Date of establishment	*Ownership*	*Location*	*Product(s)*	*Local/export*
Company 1	1943	Family business	Rural	Pickles, preserves, rose and blossom water, Arak, and other canned fruit and vegetable products	Local and export
Company 2	1990	Men owned and run	Rural	Started with a range of products, currently focusing on pickles	Currently, exclusively export
Company 3	1991	Men owned and run	Peri-urban	Rose and blossom water, traditional syrups, puddings, oils, spices, packed lentils and fruits and vegetables packed in jars	40% local, 60% export
Company 4	1978	Men owned and run	Peri-urban	Halawa tahinieh – sweet sesame paste	Export
Women's Company 5	1989	Women owned and run	Rural	Pickles, jams, molasses, syrups, rose and blossom water and several other traditional products	Local and export

cover in detail old experiences of the interviewee in order to reduce recall bias. Finally, the written transcripts allow data validation and information screening for more effective analysis and findings.

Themes

Several findings and recommendations can be drawn from these case studies. Some of these themes are common to all the case studies while others are specific to each of the interviewed companies. It should be noted that the following themes provide indications of possible strategies that the RWCs may consider to more effectively penetrate the local and international markets while taking gender issues into consideration.

Be the first in the market

Most successful companies are the ones that have been 'the first' to satisfy specific consumers' needs (Ries and Ries 2002). A company could either be the first in providing a new innovative product on the market, selling the product in a specific geographical location, or designing a new package and label. The list of being 'the first' could be extended to cover different business activities such as distribution, advertising, promotions, customer services, etc. Most importantly, companies that can be the first to satisfy the consumer's needs are more successful in retaining the loyalty of these consumers and are the ones that establish a strong brand in the market (Ries 2005).

In the traditional Lebanese food industries, there has been room for being 'the first' in different products or specific markets. The case studies show that, for instance, Company 1 was the first (or maybe the most successful among the first) to produce freshly-packed pickles. This company is currently the leader in this category of products. Also, The Women's company was the first in the market to provide products that are similar to 'home-made Mouneh' or provisions. While similar products were already on the market, they were however, produced by industries and not using traditional procedure nor by using healthy ingredients as Company 5 promises its consumers.

Being 'the first' does not only mean being the first to launch a new product in the market. It can also be the first to market the products in a specific geographical location. Company 4 was the first Lebanese company to promote its products in Arab and other foreign countries. This allowed the company to currently export about 75 per cent of its products and hence retain a strong brand in these Arab and foreign countries. Furthermore, a company can be the first in packaging design. For example, Company 5 was not only able to promote a new concept of 'home-made' products; this company was also able to provide an innovative package at a considerably lower cost. This innovation in packaging was about the sackcloth which reminds the consumers of the core values of the brand: a 'home-made' style and a healthy product. The sackcloth covered the lid of the generic container used by most industries in Lebanon and hence provided a new appearance different from the competitors (Khalil 2004).

Finally, the first kind of products in the market can generate publicity (Ries and Ries 2002). The Women's Company is a clear example of publicity where the media has been interested by the 'home-made' types of products. Publicity is an effective means of advertisement that strongly creates brand loyalty.

The RWCs should thus be able to think how their products could be 'the first' in the market. They also have to be differentiated from other industrial or 'home-made' competitors. The process and ingredients found in the RWCs are similar to other competitors: natural and traditionally processed. Hence, the ingredients and processing could not be accounted as the first in the market. If RWCs want to penetrate the market where these competitors are found, they have to focus on a new advertisement message.

This message could rely on a slogan to 'Help Rural Women's Cooperatives' located in different, underdeveloped, rural Lebanese villages. The wording of the slogan should be rephrased to be more consumer-friendly. This slogan may enforce an ethical basis for buying the items whereby consumers not only purchase a healthy product, but also contribute to the well being of these rural women.

This strategy was successfully used by many foreign companies to target young educated consumers willing to positively contribute to the society (Edwards and Day 2005, 57–71). It also follows an international trend whereby consumers are increasingly demanding upscale expensive food items produced in a healthy manner, processed traditionally and containing an ethical message (Bijman *et al.* 2006).

A different possible independent or complementary marketing message can be to develop a story for each line of the products. These stories could explain to the consumers the origins of products developed in different locations by RWCs. They could intrigue the consumers and trigger their imagination regarding each type of product. Focusing on consumers' imagination by eliciting 'mystery, sensuality and intimacy' stories can create a successful brand as argued by Roberts (2006), the CEO of one of the leading international marketing firms. Finally, designing an ethical message and creating stories for products can moreover create an emotional brand that more effectively attracts consumers (Edwards and Day 2005).

Target a narrower market

The markets have been witnessing an increase in trade liberalization whereby foreign products have become cheaper and more accessible. This has significantly increased competition where the only survivors are larger companies with leader brands or those able to reduce cost with higher volume of production. This fact has forced smaller industries to focus on niche or smaller markets not covered by these larger organizations (Schaffner *et al.* 1998).

Many successful traditional food companies in Lebanon have also targeted a narrow market. They did not launch products liked by all consumers both locally and internationally; rather they targeted specific consumers with their recipes, packaging, distribution, etc. For instance, Company 4 has effectively targeted foreign consumers willing to buy only Lebanese products. Company 2 has developed a pickle recipe only desired by Saudi consumers which is different from the Lebanese recipe. Some other companies have been even more selective and hence chosen a narrower market. Company 1 produces a Lebanese pickle recipe at a relatively high price. Only Lebanese consumers with greater means have been the target of the company. Also, the Women's Company has developed products that only target consumers willing to pay extra for a home-made style of product.

Many other food companies have tried to focus on a new target market: modern women and the younger generation (Challita 2006). The latter category is usually more educated, health conscious, searching for fun and for a positive self-image (Stefanos 2010). Hence, the RWCs may focus on younger consumers, possibly women that are more educated and searching for healthy products and a positive self-image. Knowing the exact category of consumers provides a more effective means to reach these consumers and build a stronger relationship by better responding to their needs (Perreault and McCarthy 1999).

Expand target consumer base

Successful brands do not only cover the local markets, they can also be global and reach foreign consumers (Ries and Ries 2002). The case studies have shown that there is an increasing trend in Arab and other countries to purchase products 'Made in Lebanon'. This seems to be due to two facts: the higher confidence in Lebanese products in Arab countries, and a growing global market for Middle Eastern products. For instance, Company 1 has currently received funding from foreign aid agencies to market its products to foreign consumers in the US and Canada. Company 2 believes that part of its product's success in Saudi Arabia is the fact that these products are made in Lebanon and hence are perceived of a higher quality by Saudi consumers. Company 3 is currently expanding its market due to the belief that there is an increasing demand for Lebanese products in Arab countries. The owner of Company 4 has also stressed the importance of targeting foreign consumers due to the higher demand for Lebanese cuisine in many foreign countries. The Women's Company's main consumers in foreign countries are not the Lebanese living outside Lebanon; it is rather expatriates willing to buy 'home-made' Middle Eastern or Lebanese products.

Highlighting that products are made in Lebanon provides an opportunity both in Arab countries that consider Lebanese products as higher quality and in foreign countries that are increasingly purchasing Middle Eastern products.

Innovation in product development

Product innovation is considered vital by many in the food industry in order to remain competitive (Jamrog 2006). It is an important part of a company's marketing strategy if the company aims to be 'the first' in the market. Of all the case studies, Company 1 and in part, The Women's Company are the only food plants that have launched innovative products. Though the cost of developing a new product has been high and has taken several years, especially for Company 1. This could be due to the high cost of development or to the risk of product failure (Fuller 2005). All the other new products developed in the other companies are product extensions and are only new to these companies, not to the market. This shows that new product innovation is limited in these traditional food industries, a finding that could be found in many other sub-sectors of the Lebanese food industries (MOET 2003).

Product innovation can thus be an opportunity for the RWCs where most traditional companies are unable to innovate. However, the cost of such product innovation could be high. Also, the cooperatives might not have the qualified personnel to develop such new products. There are two possible solutions for these shortcomings. First, the RWCs could rely on universities or other NGOs for product development. For example, in Latin America, many farmers have benefited from the collaboration of donors, private and public institutions to develop new Cassava products (Edwardson and Best 2008). Such collaboration is still nonexistent in Lebanon. This fact could nonetheless provide an opportunity for NGOs or other international organizations to establish a research center or simply an organization that collaborates among universities and private industries that could benefit the RWCs in product development, a finding supported by Olsen and Harmsen (2008). Since this option is currently inaccessible, the RWCs are only left with the second option which is developing a new food product themselves. Company 5 prepared new products after realizing that universities lack these services. The process of developing such products should, however, be relatively easy and cost effective since RWCs lack the finances and the qualifications to develop sophisticated food products for example as in the case of Company 1.

Innovation in packaging and labeling

Package and label innovation is one way to differentiate a product from competitors (Perreault and McCarthy 1998). Also, labeling is important to promote the identity of the product and hence should be an effective way of advertising in stores and supermarkets. Many food industries in Lebanon only differentiate their labels, rather than packaging, because of the high cost of the latter (Khalil 2004). Among the five companies that participated only the Women's Company did relatively innovate its packaging, though this innovation did not entail any change in the shape of the package. The company only relied on sackcloth to cover the white lid and the generic shape of packages used by most food industries in Lebanon. However, the company at the time was working during a war period and it is a clear example of adapting with meager resources. The glass was not judged to be attractive and the sackcloth was the only option available. All other interviewed companies relied on the generic shape of the jars and bottles, and they only differentiated their products in the labeling design.

RWCs can thus learn from the experience of Company 5 – a women run company – and try to cover the shape of the bottles and jars since designing a new package is expensive as found by other industries. This package design, along with a new label, should be able to promote the brand value of a product and reach the specific consumers for which the product is aiming. For example, if RWCs desire to target younger women, they need to develop a new label and cover the package to portray the RWCs products' brand values: healthy, hand-made, positive contribution to rural women, etc. They could also include in each label a story from the RWC that developed this product or the nature of the ingredients, etc.

Benefit from higher pricing for leader brands

Leader brands are those able to create loyal consumers by being the first to offer the product's ingredients and process, packaging, marketing slogan or others. These leader products are usually priced higher regardless of their production costs (Ries and Ries 2002). When products are not market leaders, they are then similar to other competitors and in this case they have to lower their prices to effectively penetrate markets (Perreault and McCarthy 1998). For example, Company 1 was the first to launch the 'freshly packed' pickles. The prices of its products remained high despite the entry of other competing brands. Company 4 was 'the first' to introduce its products in Arab countries and its prices remained high in the markets. Company 5 is the leader in 'home-made' products since it was 'the first' to introduce this range of products to consumers. The prices of The Women's Company are still high despite the fact that other 'home-made' products have been introduced into the market with lower prices.

Other companies that are not leaders in the market are competing with lower prices. For instance, Company 3 is expanding its production capabilities to reduce the cost and the prices of products. This is due to the fact that the products that will be launched are not new to the market and hence have several competitors. Company 2 on the other hand, has followed both strategies. To be able to penetrate the Saudi market the company reduced its product's price and successfully competed against the only other brand of pickles, produced in Greece. The company has won leadership after gaining most of the Saudi consumers. After this leadership has been established, the brand's price was not reduced even after the introduction of competing brands with lower prices.

The prices of the RWC products greatly depend on the ability to be 'the first' in the market. If RWCs are successful as market leaders, then their prices could be high, even higher than the other competing home-made and industrial products. However, if the RWCs are unable to differentiate their products and gain leadership, then their prices should be competitive. They might even have to consider setting lower prices than other home-made products due to similarity in ingredients and production techniques. They can nonetheless keep the prices higher than industrial products due to use of natural ingredients and traditional type of processing.

Advertise through direct contacts and e-social networks

Advertising is the means through which consumers can learn about a specific product. The channels through which Arab companies advertised their products have changed where currently many rely on direct and internet advertisements (Henneberg, Naudé and Mouzas 2010). None of the companies interviewed used the traditional means of advertising (TV and radio) since their cost is high for SMEs and their possible return to investment is low. These companies rather prefer to rely on labeling, promotions and exhibitions as their only means to advertise their products. For example, Company 4 has relied on foreign exhibitions to market its products and regularly offers product or price promotions. Also, The Women's Company relies on packaging, exhibitions and promotions to attract old and new consumers.

RWCs can design a new package to effectively promote the brand's values in supermarkets. They should also assist in exhibitions to target new potential consumers and agents willing to buy such products. Designing labels and attending exhibitions are both considered a type of direct advertisement. Consumers could be informed about the product values from the label's information on supermarket shelves. Also, exhibitions could allow the consumers to taste the products and be informed, face-to-face, on the possible brands' benefit. RWCs might furthermore rely on a private marketing agency or a marketing consultant to build an internet social networks such as facebook and twitter to target the younger generation.

Find an agent for distribution and use sales representatives

The case studies show that all the companies have relied on an agent for distribution. This agent is either local or international, and the company could either totally or partly rely on this agent to cover the local market. For instance, Company 3 and Company 4 have both their own local distribution channels and local private agents for different regions in Lebanon. The agents usually cover regions not targeted by the companies' own distribution chains. On the other hand, the other companies completely rely on private agents. Company 1 and The Women's Company have both preferred to have a private agent who is responsible for distributing the products and collecting payments where the latter task is considered cumbersome by these companies.

All the interviewed companies have commissioned distributors in external markets. Building a distribution network in foreign countries is costly due to the high investment needed and to unfamiliarity with each of the targeted foreign markets. However, a couple of companies invested in travelling and in personal relationships.

Finding local and foreign distributors can only be achieved by the RWCs when a narrow market for the products is defined and packaging specifically designed to target this market. Distributors are usually interested by brands demanded by consumers that eventually increase profits. These distributors might sometimes direct the companies on which kind of promotions could be achieved to further increase sales as is the case of The Women's Company. However, the role of the distributors remains limited if the RWCs lack a well defined marketing strategy. Hence, it is imperative that RWCs focus first on the targeted market, quality, packaging, and a strategy for advertisements before assigning local or foreign agents.

Keep an eye on quality and consistency

Quality is one of the main focuses of the food industry. Producers must concentrate on quality in order to remain competitive in the market. While defining quality relies mainly on the consumer perception of quality (Vasconcellos 2004). For this reason, the case studies can provide a definition of what is considered high quality for the traditional food products. Company 1 and the Women's Company considered quality as any product that is 100 per cent natural and free from chemical additives. Company 2, Company 3 and Company 4 argued that products 'made in Lebanon' are considered 'higher' quality by many Arab consumers due possibly to the perceived good taste of these products. Moreover, most of the companies interviewed have stressed the importance of keeping the recipes consistent and having a 'tasty' product.

A broad definition taken from the case studies to mark quality in traditional food products can be the following: products that are 100 per cent natural (free from preservatives and chemical additives), tasty with consistent recipes and finally 'produced in Lebanon' (to target Arab consumers which consider Lebanese products as high quality). RWCs products are all 100 per cent natural which partly satisfies the definition of quality. Also, these products are all produced in Lebanon and hence, the RWCs should stress on the front labels their place of production if the products are aimed at Arab consumers. Yet, RWCs may face problems of lack of consistency as stated by some of the interviewed companies. Inconsistent recipes may also affect what is considered as 'tasty' by consumers where recipes have different taste and colors in every batch. Producing consistent products entails training and special attention by all employees working in cooperatives to constantly monitor quality throughout the production chain.

Quality control certifications – ISO is the way of the future

One way to achieve consistency is through implementing structured quality assurance programs and acquiring international certifications of quality like the ISO 22000. Company 3 was in the process of taking the HACCP certification and possibly the ISO in the future to be able to adequately control quality.

Other companies interviewed were not enthusiastic regarding quality control programs and certifications. For instance, Company 1 finds that certifications will only be important for the future when they will be required by foreign markets and governments. Company 2 and The Women's Company do not believe in the importance of such certifications to boost their sales. They rely on their Food Products Association (FPA) certification. More importantly, Company 4 has already an ISO certification where the owner doubts its positive contribution to sales and to the company. This is due to the high cost of certifying the plant and to the nature of these traditional products that possibly do not require such rigid quality control programs.

Despite their cost, these certifications guarantee the quality of products and could greatly facilitate the products' entry to market if foreign governments impose new regulations on Middle Eastern traditional items.

RWCs may implement an informal quality control program without currently seeking costly certifications. Such a program should be coupled by a managerial culture whereby quality should be considered the top priority of employees (Borch and Roaldsen 2009). The program would facilitate the future implementation of international quality control programs and certifications like the ISO and HACCP if these certifications become a requirement in local or foreign markets.

Strong personal initiatives and good interpersonal relationships

Interpersonal relationships are found to be important in these case studies. Company 2 has stressed the importance of strong relations with foreign agents to help reduce competition. Company 4 has also emphasized the importance of personal relationships in order to increase market share. Strong relationships are due to an effective customer service or to building friendship that boosts trust especially in Arab countries where such relationships might be considered important.

Furthermore, personal initiatives and motivation are found to be part of the success. Personal initiative could offset any lack of government assistance to the food industry and to the countries' political and economic situations. For example, the owner of Company 4 has stressed the importance of personal initiatives to attend foreign exhibitions that are not financially supported by the Lebanese government. The Women's Company has emphasized the risk taking role played by the owners during the civil war to start the new business.

RWCs should thus expect higher personal involvement if they need to further expand their markets. One possible solution is to form a committee that could address their general problems and help them access markets as the food industries have done in establishing their own syndicate. The latter has allowed the industries, among others, to more effectively participate in foreign exhibitions. This RWC joint committee could collaborate with the syndicate of the food industries and learn from their experience to effectively target new markets.

Extend the markets and product lines progressively

If the RWCs need to increase their market in supermarkets and foreign countries, they may have to learn from the experience of the interviewed companies. These companies have started with a small market and then progressively expanded to others. For example, Company 1 targeted specific regions in Lebanon then moved to all other regions and is currently seeking to target foreigners. Company 4 began through exhibitions to target foreign markets and currently it is able to export products to the five continents. The Women's Company began with only one supermarket in Lebanon and then expanded to others after the brand's success with consumers.

If RWCs plan to cover supermarkets both locally and in foreign countries, they could start with one market and with a limited range of products. For example, Company 4 started with tahini and halawa, then developed many other items. The Women's Company also started with just a few products in one supermarket then expanded to others. Company 2 is successfully marketing only one product to only one market and is currently planning to extend its product line and its market.

Moving progressively in the market and with product lines has several benefits. First, it reduces the cost of failure especially if the products have not been present in the market and have not yet established a strong link with the consumers. Second, it allows a more effective production where a larger volume of products could be beyond the production capabilities of RWCs. Finally, it allows a better control of quality whereby each cooperative can be specialized in one range of products. This allows more effective monitoring and verification of the whole process of production.

Apply the law of 'focus' when possible

Most successful companies have focused on a few products or target markets in order to effectively compete (Ries 2005). For example, Company 2 has only one product and focuses on only one market. Company 1 and Company 4 have both focused on a narrow market then moved to cover other markets. The Women's Company targets only one type of consumer; it started locally then extended globally.

If the RWCs want to gradually expand their markets and product lines, and if they plan to target a narrow market as discussed above, then they may simply apply the law of 'focus'. Furthermore, since several cooperatives produce different types of food items, each cooperative might have to specialize in only one product to cover the target market. Specialization might improve the production capability and quality control of each of the cooperatives.

Building bridges with women's rural cooperatives

Since the RWCs partly produce the same range of items as the traditional food manufacturers, collaboration between the RWCs and the food industry could for this reason be positive in many ways. A food company unable to produce certain types of product might assign the production task to the cooperatives. Also, a food industry might be interested in marketing some or all of the cooperatives products if they are found to increase the company's profit.

In the case studies, the interviewed companies were thus asked about their opinions on the RWCs, and whether they have worked or would be interested to work in the future with these cooperatives. All the companies were found to not be currently working with the cooperatives. Their opinion on such a relationship and the possibility of future collaboration, however, differed from one case study to the other. The following summarizes the general responses of the interviewed companies.

At Company 1 the owner explained that the company has previously worked with the RWCs but found their products expensive despite the fact that the RWCs 'produce the same items as the company does'. Also, the owner believes that the RWCs have problems with product consistency where products' taste and color change from one batch to another. The owner expressed the possibility of joint collaboration, especially in products like 'fig jams, date and grape molasses'.

Company 2: the owner has no plans to work with the RWCs and is not completely informed regarding their work and products.

Company 3: the general manager has worked with the RWCs in his previous job where the relationship was positive. The task of the RWCs was mainly manual and related to chocolate packaging. He believes that they could be of strong help especially in products that the companies might produce in the future.

Company 4: the owner has previously worked with a charity organization where he marketed its products in foreign exhibitions. The owner could be interested in marketing products made by the RWCs and not produced by his company.

The Women's Company: the owner believes that these cooperatives might have problems of inconsistency since each product is produced by different RWCs, hence each product might have a different recipe and taste. The owner does not plan to work in the future with the RWCs.

In conclusion, some food industries might use the help of RWCs especially in hand-crafted products as in the case of Company 3. Some others might be interested in marketing their products as the case of Company 4. However, the RWCs should consider maintaining consistent recipes through adequate monitoring of quality. This will allow them to more effectively collaborate with industries like Company 1 that is interested in some of the products but fear inconsistencies in recipes.

Conclusion

In terms of operation, an already successful enterprise is gender blind. The enterprise has broken the gender barrier. However, the challenge remains for organizations such as RWCs that are carving their niches in a highly competitive environment. For them, gender-related factors are both enabling and disabling.

The enabling dimension of the gender factor is the tenacity of women and their skills that are compatible with the demands of the food industry. That was stated by all respondents from the five companies. Another enabling dimension is the level of empowerment which has the potential to create a will to unleash the all powerful entrepreneurial current that has the capacity to tear down barriers and survive failures; The Women's company is a testimony to that fact. The founders worked during a period of high political instability and risked their lives to obtain the necessary resources. They managed to negotiate access to financial resources in a highly patriarchal environment.

Factors working against successful operation constitute the invisible barriers of patriarchy in various accents that are prevalent in the Arab region and globally that limit women's access to financial resources and women's networking capabilities. However, local social barriers to women's work, especially in rural areas, are receding in view of financial challenges. One indication is the willingness of parents from the highly conservative Akkar area to allow their daughters to stay outside the home.

The evidence provided by the case studies of five small establishments indicates the necessity to conduct more intensive research that integrates, more systematically, gender issues in small businesses in Lebanon.

Notes

1 The Collective for Research and Training on Development – Action (CRTD.A).
2 University of Balamand (based on initial papers by Tonnie Choueiri and Raymond Kasatly).
3 Kalleberg and Leicht 1991, p. 160
4 Kalleberg and Leicht 1991, p. 160.

5 http://slfi.org.lb/

6 The size of the industry is defined according to its number of employees; a small industry employs between 10 and 50 employees while a medium industry employs between 50 and 250 employees. There are several definitions of SMEs. The number of employees differs according to each definition. In the US, the number of employees in the small and medium enterprises is less than 100 and 500 respectively. This study relies on the EC definition since it better reflects the nature of the Lebanese market which is smaller in size than the American market; http://ec.europa.eu/enterprise/policies/sme/facts-figures-analysis/sme-definition/index_en.htm

References

Bijman, J., Omta, S.W.F, Trienekens, J.H., Wijnands, J.H.M and Wubben, E.M.F. (2006) 'Management and Organization in International Agri-Food Chains and Networks'. In Bijman J. *et al.* (eds) *International Agri-food Chains and Networks.* Wageningen Academic Publishers, the Netherlands.

Bird, S.R. and Sapp, S.G. (2004) 'Understanding the Gender Gap in Small Business Success, Urban and Rural Comparisons', *Gender and Society* 18(1): 5–28.

Borch, O.J. and Roaldsen, I.H.E. (2009) 'Competitive Positioning and Value Chain Configuration in International Markets for Traditional Food Specialties'. In Canavari, M. *et al.* (eds) *International Marketing and Trade of Quality Food Product.* Wageningen Academic Publishers, the Netherlands, p. 347.

Challita, D. (2006) 'New Marketing Strategies'. *Lebanon Opportunities,* (July): 29–30.

Edwards, H. and Day, D. (2005) *Creating Passion Brands.* Kogan Page: London, p. 244.

Edwardson, B. and Best, R. (2008) 'Rural Agroenterprises: Cassava development in Latin America'. In Earle, M. (ed.) *Case Studies in Food Product Development.* CRC Press: Boca Raton.

Fuller, G.W. (2005) *New Food Product Development: From Concept to Marketplace.* CRC Press: Florida.

Jamrog, J.J. (2006) *The Quest for Innovation: A Global Study of Innovation Management 2005–2006.* Human Resource Institute, University of Tampa: Tampa, FL.

Khalil, R. (2004) Packaging: More Than Graphics. *Arab Ads* (June): 94–95.

Kalleberg, A. L. and Leicht, K. T. (1991) 'Gender and Organizational Performance: Determinants of Small Business Survival and Success', *Academy of Management Journal* 31(1): 136–61.

MOET (2003) 'Competition in the Lebanese Economy: A Background Report for a Competition Law for Lebanon'. Ministry of Economy and Trade: Beirut, Lebanon.

Olsen, J.R. and Harmsen, H. (2008) 'Product Development Alliances: Factors Influencing Formation and Success'. *British Food Journal* 110,(4/5): 430–43.

Perreault, W.D. and McCarthy, E.J. (1999) *Basic Marketing: A Global Managerial Approach* 13th edn. Irwin/McGraw-Hill: New York.

Ries, A. (2005) *Focus: The Future of Your Company Depends on It.* Harper Paperbacks: New York.

Ries, A. and Ries, L. (2002) *The 22 Immutable Laws of Branding.* HarperCollins: New York.

Roberts, K. (2006) *The Lovemarks Effect: Winning in the Consumer Revolution.* PowerHouse Books: Brooklyn, NY.

Schaffner, D.J., Schroder, W.R. and Earle, M.D. (1998) *Food Marketing: An International Perspective.* McGraw Hill.

UNIDO (2009) *Industrial Development Report 2009. Breaking In and Moving Up: New Industrial Challenges for the Bottom Billion and the Middle-Income Countries*. UNIDO: Vienna.

Vasconcellos, J. A. (2004) *Quality Assurance for the Food Industry*. CRC Press: Florida.

Henneberg, S.C., Naudé, P. and Mouzas, S. (2010) 'Sense-making and Management in Business Networks: Some Observations, Considerations, and a Research agenda'. *Industrial Marketing Management* 39(3): 355–60.

8 Gender inequality in the context of climate change

The case of the Boudinar commune in Morocco

Abdellatif Khattabi[1], Manar Matah[2] and Soumaya Ibrahim Huber[3]

Introduction

Women's and men's differential access to social and physical goods or resources is one of the key dimensions of gender inequality. This paper considers some of the ways in which these inequalities are deepened by the impacts of climate change, and can be seen as a pre-warning that gender inequality if not addressed properly can limit the effectiveness of climate change adaptation strategies or any developmental efforts.

Earlier studies related to the impacts of climate change in the commune of Boudinar showed that this area is prone to several climatic events which have affected the living conditions of the local population. The rural commune of Boudinar and the whole Oriental region, located in north east Morocco, has an arid and semi-arid climate. This commune is located on the coastal zone of the province of Driouech, and rain fed agriculture ('Bour') is important in the local economy contributing, in addition to animal breeding, to 75 per cent of the local income (Ez-Zaher 2008). However, water resources there are threatened by over-use, pollution and intrusion of sea water into aquifers.

The climatic and hydrological context of the rural commune of Boudinar is very sensitive, because of the irregularities of rainfall. During the 80s and 90s, the area witnessed periods of successive droughts which aggravated the poverty, as it led to the degradation of land and losses of agricultural production (Faouzi and Khattabi 2009c). Moreover, the analysis of the climate observations recorded in the area of the Eastern Mediterranean coastal region of Morocco (DMN 2007) have shown future projections of evolution of the climate with tendencies towards an increase in the temperatures, a reduction in rainfall, and an increase in the frequency and length of droughts accentuating imbalance between supply and demand of water.

As the commune of Boudinar is located in coastal zones, its aquifers are subject to sea water intrusion. Several problems related directly to water resources are felt in the commune. Among the consequences of the climate change impacts, the water scarcity constitutes an important issue. The water scarcity has significant impacts on agricultural activities and on potable water availability. The local economy which is primarily based on traditional agriculture and extensive animal breeding is negatively impacted.

The lack of sufficient wells and of water storage tanks makes it difficult to fetch water (Faouzi and Khattabi 2009b). The social layers most affected by the water scarcity, induced by climate change, are women and young girls who are the main group in charge of water collection from distant wells and springs. The vulnerability of the female community members is aggravated by their social situation, limiting their mobility and access to and control of resources (Faouzi and Khattabi 2009b).

It is for all these reasons that this research is focused on the commune of Boudinar, its physical resources and its living conditions in order to raise awareness and search for potential adaptation solutions suitable for both men and women that address water shortages induced by the changing climatic conditions. This chapter deals with some of the gaps in understanding how men and women farmers cope with climate variability. It explores how coping strategies are shaped by gender dimensions of power relations and the decision-making process.

Brief overview of the current knowledge and past research on the problem

Though the issue of climate change is not new, more recent is the integration of a gender-sensitive perspective in climate change research and responses. Even more recent is the integration of a gender-sensitive perspective in climate change research and responses. For this reason, there is little existing research considering the linkages between climate change and gender. Similarly, while there is a wealth of literature on gender and the environment, gender and energy, gender and water, gender and conflict and gender and disasters, there are few explicit references to gender and climate change (Bridge 2008).

Climate variability and climate change

Climate variability refers to variations in the mean state of the climate and variations, such as the occurrence of extremes on all temporal and spatial scales beyond that of individual weather events. The average range of temperature for a location, as indicated by minimum, maximum and average temperature values, is an example of a measure of climate variability. This differs from climate change which refers to a long-term change in the state of the climate and which is identified by changes in the means and/or changes in the variability, or changes in the frequencies or intensities of extreme events.

This chapter describes climate variability at different points in time based mainly on farmers' descriptions of 'average weather'. Farmers' descriptions are given equal footing with scientific measurements because documenting how farmers frame the conditions they experience is crucial for understanding their responses to said conditions (Roncoli 2006; Tschakert 2007). This work refers to changes in climate variability without necessarily concluding that a permanent change has taken place (a state that would be referred to as climate change) (FAO 2010).

Coping and adapting

This study adopts the terms as coined by Davies (1993), who distinguishes between farmers' short-term and long-term responses to climate variability and food insecurity. It defines short-term responses to a decline in food availability and income in abnormal years as coping or coping strategies, and defines longer-term or permanent changes in the ways in which food and income are acquired as adapting or adaptation strategies.

Climate change is a global phenomenon, and its impacts are already being experienced on a human level. It is recognised that it is those who are already the most vulnerable and marginalised who experience the greatest impacts (IPCC 2007), and are in the greatest need of adaptation strategies in the face of shifts in weather patterns and resulting environmental phenomena. At the same time, it is the vulnerable and marginalised who have the least capacity or opportunity to prepare for the impacts of a changing climate or to participate in negotiations on mitigation. As women constitute the largest percentage of the world's poorest people, they are most affected by these changes. Children and youths – especially girls – and elderly women, are also highly vulnerable. Even where there is a lack of hard evidence, it is commonly recognized that climate change exacerbates existing inequalities in the key dimensions that are not only the building blocks of livelihoods, but are also crucial for coping with change, including: wealth; access to and understanding of technologies; education; access to information; and access to resources (Masika 2002). But what does this mean in practice?

A gender-sensitive response requires more than a set of disaggregated data to show that climate change has differential impacts on women and men. It requires an understanding of existing inequalities between women and men, and of the ways in which climate change can exacerbate these inequalities. Conversely, it also requires an understanding of the ways in which these inequalities can intensify the impacts of climate change for all individuals and communities. For example, men may have greater access to vital information on climate change mitigation or adaptation strategies for cultural reasons, or because women are too busy with caring and other domestic responsibilities. This lack of information and lack of opportunity to feed their own knowledge into community or national-level adaptation and mitigation strategies could jeopardize larger processes of reducing climate change and its impacts.

Gender sensitivity in consultation and decision-making is also essential for effective mitigation and adaptation responses to climate change. More than simply thinking about how these processes can be tailored to the specific needs of poor and vulnerable men and women, there is a need to recognize the capacity of women and men, girls and boys, to contribute important knowledge and insights. With more participative processes, these strategies and interventions can truly identify and meet the needs of those they aim to assist. In this way, processes can be forged that respond to local realities while feeding into a broader vision of climate change deceleration. Yet women are more likely than men to be absent from decision-making, whether in the household or at community, national or international levels – either because their contribution is not valued or because they do not have the time, confidence or resources to contribute. Recent research by the Institute of Development Studies and Plan International has also pointed to the marginalization of children's voices in household, community and national decision-making relating to climate change – particularly in disaster risk reduction (Mitchell *et al.* 2008). It is critical that more is done to promote women's and children's meaningful participation in decision-making on climate change responses, to ensure that climate change policy and grassroots interventions respond to their specific needs and draw on their knowledge and experience. In this way the profile and status of women and girls in the community can also be raised, while challenging traditional assumptions about their capabilities (Bridge 2008).

Research objectives

This research tests the hypothesis that: 'because of the unequal gender roles (the behaviours, tasks, and responsibilities a society defines as "male" or "female") and the differential gendered access to resources, men and women experience climate variability differently and cope in unequal ways with climate variability'. Gender inequality is thus a major factor contributing to the increased vulnerability of women and girls in disaster situations. These inequalities in existing gender relations and the social expectations influencing them need to be researched and change needs to happen. Accordingly the aim of this study is to understand how men and women in farm households in Boudinar perceive and experience climatic shifts. The main objectives of this research were to:

- identify the social context in Boudinar and how it relates to the physical context and the local climate conditions and risks and to identify trends in climate variability over the past decades as perceived and experienced by men and women farmers;
- understand how men and women in farm households perceive and experience climatic shifts and how this is linked to their livelihoods and food security;
- identify the coping strategies that men and women farmers utilize in order to ensure their livelihoods in response to climate variability and to understand the resources and decision making processes utilized, and to assess the related outcomes.

The importance of the research questions lies in their development implications, because this study with its findings can be seen as a sample of a research exercise that could be repeated elsewhere. Further on it provides a contribution which could portray enough convincing evidence to justify any interventions or infrastructural costs involved.

The socioeconomics of the Boudinar commune

Administrative structure

The rural commune of Boudinar is located in the coastal zones of Driouech province in the Oriental region of Morocco. Created in 1958, the commune of Boudinar belongs to the Caïda of Temsamane, Circle of Rif. It has a population of 10,504 inhabitants (HCP 2004). It has a land area of about 8,328ha including 60ha of forest (Rifai 2007). It is bounded in the north by the Mediterranean Sea, in the North-west by the commune of Oulad Amghar, in the North-east by the commune of Beni Marghine, in the South and South-west by the commune of Temsamane and by the two communes of Oulad Amghar and throughout in the west (Figure 8.1).

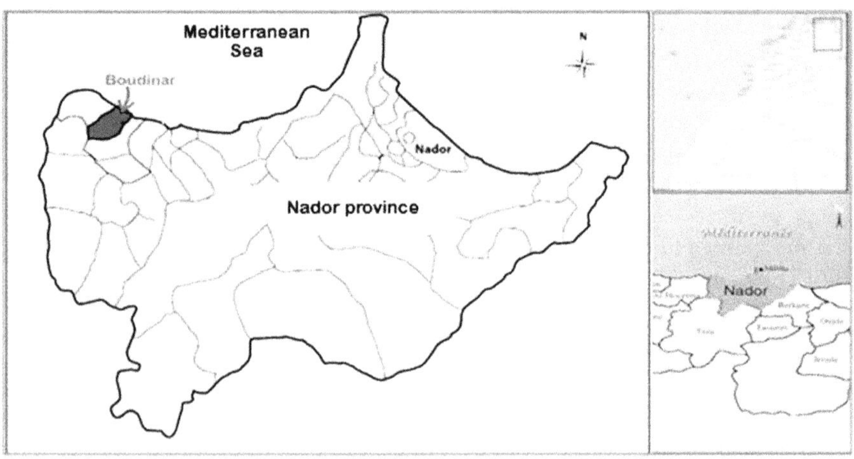

Figure 8.1 Geographical location of the commune of Boudinar: the study site

Land morphology, geology and soil types

Almost the entire territory of the commune of Boudinar (70 per cent) consists of eroded lands. The commune is part of the Boudinar basin which is located to the east of Al Hoceima bay in an area of transition between Central Rif, high and humid, and the vast arid basins and plains which extend to the East, of triangular form, largely open to the sea where it borders for 20km (Couprie 1998).

It is a relatively low area not exceeding 300 to 400m in altitude, surrounded in the South by the mountain of Beni Touzine, in the East by the mountain of Beni Saïd and in the West by the mountain of Trougout. The heights in these mountains exceed 1100m in Trougout, 1000m in Beni Touzine and 990m in Beni Saïd (Couprie 1998).

Climate

The local climate is Mediterranean and arid with 79 per cent of the rain falling between November and May. The annual average of precipitations is 250mm. The number of days of rain does not exceed 46 days per annum with an inter-annual irregularity of precipitations. In the last 25 years, data indicate that the most rainfall occurs in November, whereas in previous years, it normally occurred in April and January (Rifai 2007). The winter corresponds to the rainiest season and the summer corresponds to the driest season. According to Couprie (1998), the seasonal thermal variations are very important. The maximum temperatures are recorded in summer (23.7C in July and 24.1C in August) and the minimal temperatures are recorded in winter (11.9C in January). The annual temperature average is of 17.4C, which indicates a mild climate. The average solar hours approximates 3000 hours per annum with 300 in July.

The province of Driouech to which the commune of Boudinar belongs is dominated by the western and south-west winds which blow in winter and the east-north-east winds which dominate in summer. The average annual speed of these winds is 4.3m/s. It can, however, increase in a considerable way especially in winter for the north-western and north-north-west winds to reach values exceeding 11m/s. The 'bioclimate' of an area is the result of the interaction between several climatic factors (rain fall, temperatures, hygrometry of the air, etc). Analysis of the bioclimate, through the coefficient of Emberger and the average minimal temperature of the coldest month, shows that the bioclimate of the commune of Boudinar is mainly semi-arid.

Cropping patterns (Rifai 2007)

Agricultural land covers the largest part of the commune territory with about 4000ha, that is roughly 46per cent of the total surface. The fallow land which is much eroded occupies a very important proportion with 3,200ha. The river bed 'Oued Amekrane' lies on 624ha, that is 7 per cent of the total surface. The coniferous reforestation represented mainly by the plantations of Pinus halepensis in the mountain of Dhar Oubarren, as well as in the borders of Oued Amekrane, covers a surface of 59ha, that is 0.7 per cent of the total surface (Figure 8.2). The region suffers from major problems due to erosion and flooding during rainy periods, which destroys the facilities through water flows and the Amekrane river overflow.

Hydrology and water availability and use in Boudinar

The hydrographic network in the commune of Boudinar is centered on the Amekrane River which crosses the commune from one end to another with a large bed width (Rifai 2007). It has a length of 35km, an annual medium flow of 2.2 m^3/s, a coefficient of streaming of 0.40, a specific flow of 6.5$l/s/km^2$ and a medium flow in the driest month of 0.5m^3/s (Sabhi 1998). With its effluents, it drains the main part of surface waters of the basin. During the summer season,

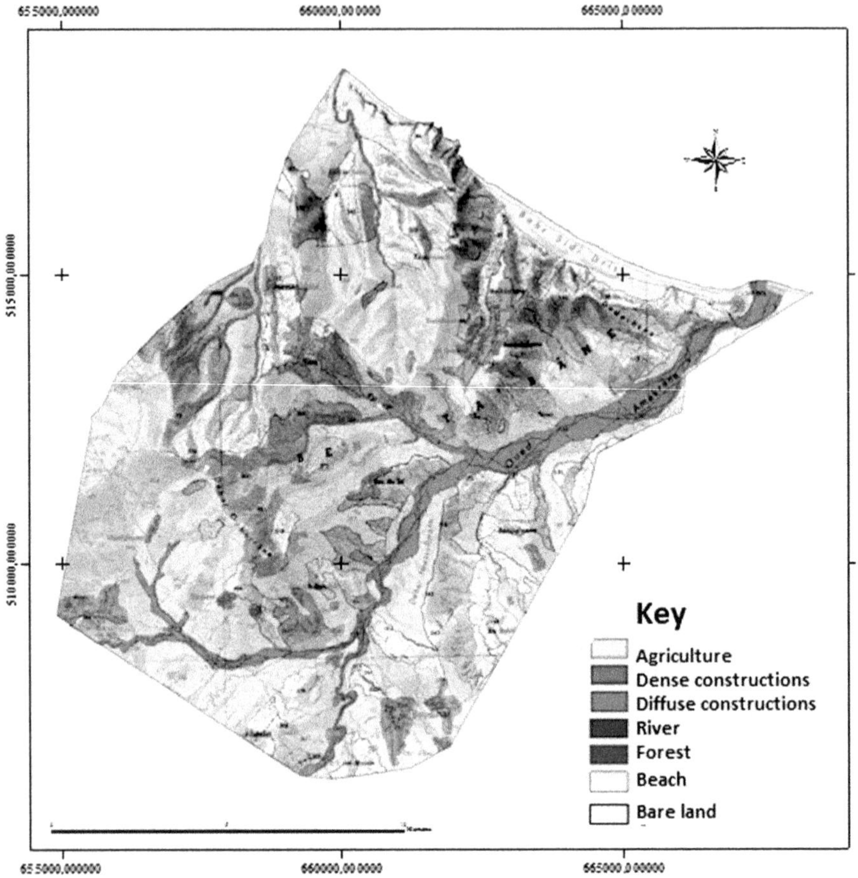

Figure 8.2 Land occupation of the commune of Boudinar

Source: adapted from Rifai 2007

water rarely reaches the sea because of its diversion for irrigation of agricultural lands located on the river edges.

The total water sources in the commune of Boudinar were estimated to 252, including 59 water springs and 193 wells. Furthermore, there are also seven water storage lakes (four in the village of Ijtti, one in the village of Saida and two in the village of Tizza). The water collected from these sources is used by the population for daily needs (consumption, irrigation, domestic tasks) and for watering animals. However, only a small proportion of these water sources is still used for supplying potable water.

There are four types of water source in the commune, namely wells, springs, a river and lakes used to collect rainfall. Inhabitants, who cannot cover their water needs from the nearby water sources, resort inevitably to purchasing it from big wells built by some well-to-do families and store it in built-in tanks at home.

The spatial distribution of water sources in each village by cluster

These water sources are unevenly distributed in the villages of the commune (Figure 8.3).

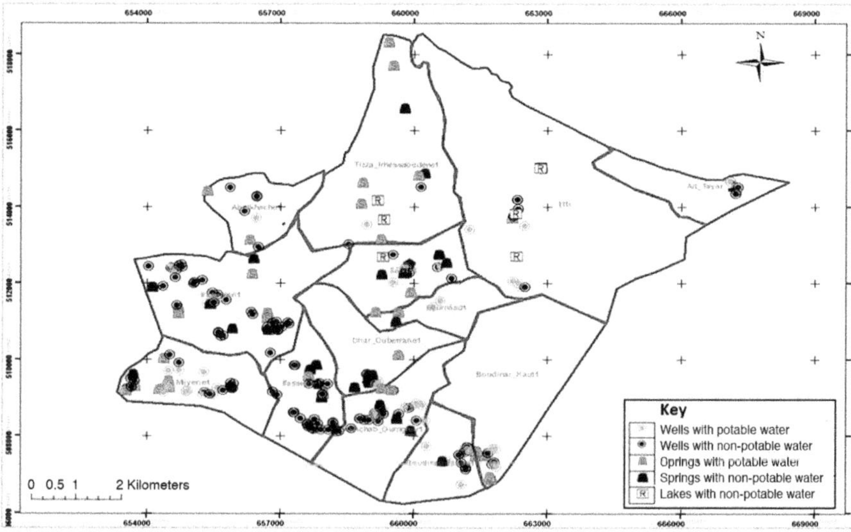

Figure 8.3 Distribution of water sources (wells and springs) in the Boudinar commune

Among all villages, Irchamene contains the largest number of water points (53), the largest number of wells (44) and many water springs (9). Contrary to this, the village of Boumâad contains the smallest number of water springs (5) and the smallest number of wells (3) as well as Douar Ait Tayar, Tizza which does not contain any water springs (Figure 8.4).

From 191 functional water sources, 59 per cent are used for water supply for consumption while 41 per cent are used only for supplying water for irrigation, cattle watering, and the domestic tasks (Table 8.1).

Table 8.1 Water sources by states of water availability and quality in Boudinar

State of water source			*Type of water source* *Springs*	*Wells*	*Total*	
Functional	Consumable water		36	77	113	191
	Non consumable water	Salted water	5	47	52	
		Others	8	18	26	
Abandoned	Dry		5	6	11	63
	Others		5	47	52	
Total			59	195	254	

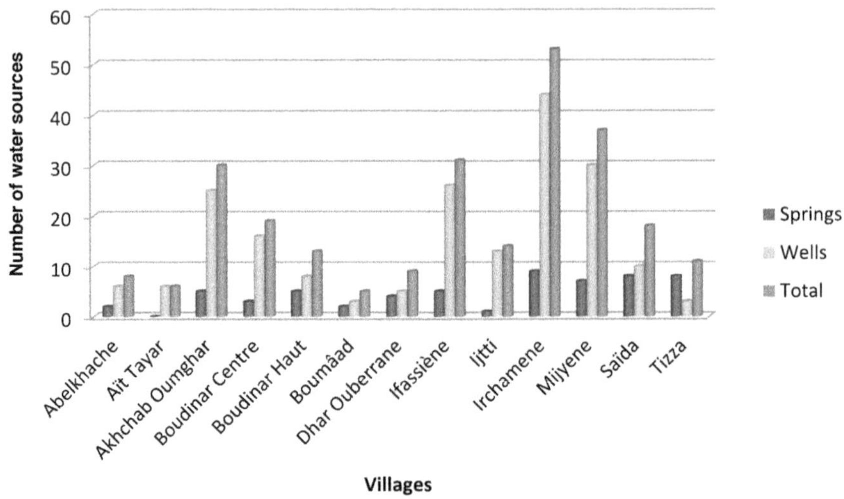

Figure 8.4 Number of water sources in the Boudinar commune, by type and village

Besides, in each village connected to the network of water, certain points, containing some appropriate water for consumption, are used for irrigation or for domestic uses.

There are 52 water sources, (47 wells and five springs), that have been abandoned because of their low water quality. There are 78 functional water sources which are not used for potable water because of their degraded water quality but they are used for other purposes (domestic uses). The water is either salty (sea water intrusion into aquifers), or polluted. There are certain zones that are more affected by the water degradation than others and the salinity of water sources is not equally distributed over the commune. The problem of water salinity is more frequent in the villages of Ifassiène, Akhchab Oumghar, Dhar Ouberrane and Irchamene (Figure 8.5) whose percentages of points with salted water are respectively 67.7 per cent, 36.7 per cent, 33.3 per cent, and 22.6 per cent.

In addition to the salty water sources, which are still used by the population for various purposes, 25 per cent of the water sources or 63 points (10 springs and 53 wells) are unused because of drying up, destruction by floods or landslides or difficulty with access. It should be noted that the water points which contain saline water are inevitably used by the people of the villages for different purposes (cleaning, watering of livestock,) however, the problem of complete abandonment of water sources is found in almost all villages. In both the villages of Abelkhache and Aït Tayar, half of their water sources have been definitively abandoned and are not used anymore. The other villages, which face the same problem are, in the following order, Irchamene, Ijtti, Dhar Ouberrane, Saïda, Mijyene, Akhchab Oumghar and Ifassiène (Table 8.2).

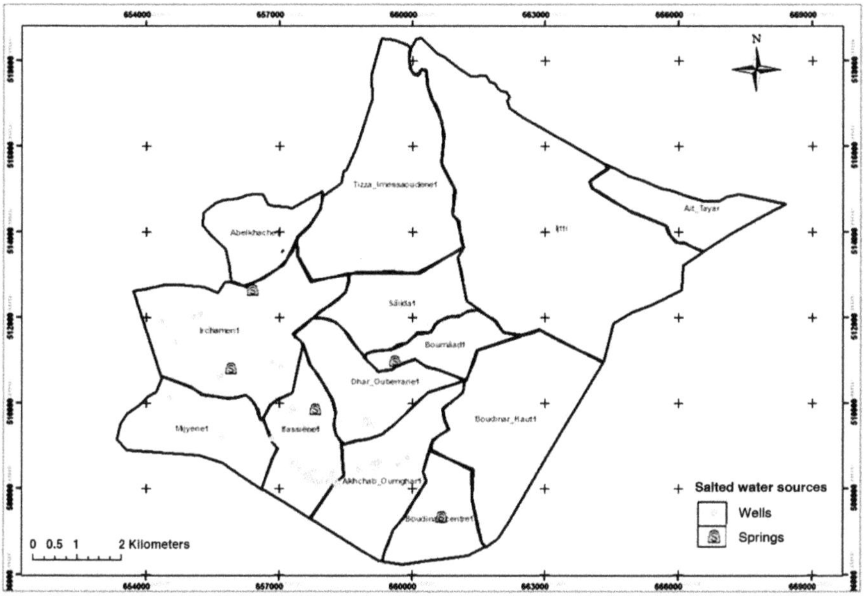

Figure 8.5 Distribution of points with salt water on the territory of Boudinar

Some villages of the commune are connected to piped, potable water. The state of this network rather is not satisfactory given the many breakdowns which arise all year round. In addition, the connection rate to drinking water supply is only 34 per cent. Other villages, not connected to the network, obtain their drinking water from wells and natural springs. Thus, the shortage of drinking water poses a heavy constraint on the lives of local people.

Table 8.2 Water sources abandoned by village of the rural commune of Boudinar

Village	*Abandoned springs*	*Abandoned wells*	*Total of abandoned water points*	*Percentage of abandoned water points (%)*
Abelkhache	–	4	4	50.00
Aït Tayar	–	3	3	50.00
Akhchab Oumghar	1	5	6	20.00
Boudinar Centre	–	1	1	5.27
Boudinar Haut	–	1	1	7.70
Boumâad	–	–	0	0.00
Dhar Ouberrane	3	–	3	33.33
Ifassiène	–	5	5	16.13
Ijtti	–	5	5	35.74
Irchamene	2	17	19	35.85
Mijyene		9	9	24.32
Sâïida	3	3	6	33.33
Tizza	1	–	1	9.10

At Boudinar centre, which is already connected to the network of piped potable water, some peripheral parts were not linked with this network. The project to build a network of piped potable water will certainly ease the accessibility to the water for the populations and will help decrease the chores of water collection. To strengthen this network of potable water infrastructure in the rural commune of Boudinar, some projects were implemented and others are in progress and will benefit the population of some villages. Indeed, the projects '*Supply of villages of Ijetti, Saïda, Ait Tayer and Ifassiene of piped potable water*' are already in progress and will allow the populations of these villages to have access more easily to potable water. However, there are many villages, with rather big populations, which will continue to suffer because of the scanty water resources and will rely on water collected from wells and springs.

Infrastructure

The road network is not fully developed within the commune. But there has been a significant development: the construction of the new coastal road linking the Mediterranean cities Oujda and Tangier. Given the rugged and sloping topography of the commune, the road network remains very limited and most of it is not paved and not easy to use during the rainy season. However, since the construction of the coastal road in summer 2006 which crosses the commune along 9km, some villages (Ait Tayar, Ijtti; Tizza, Abelkhache, Boudinar centre and Boudinar top) have benefited from the new ease of communication and transport.

The infrastructures of liquid and solid waste treatment and disposal are almost absent. For households' solid waste, it is often the population itself who takes care to get rid of it. In fact, these wastes are discharged into the surrounding environment without any prior treatment (Anonymous 2010).

The rate of connection to the electricity grid is 96.8 per cent. The commune has basic schools (primary and secondary schools). The number of doctors and health professionals falls far short of needs. In addition, the problem of remoteness and lack of roads accentuate the isolation.

The Boudinar population

The inhabitants live in the 13 villages of the municipality, namely: Abelkhache, Aït Tayar, Akhchab Oumghar, Boudinar centre, Boudinar Haut, Boumâad, Dhar Ouberrane, Ifassiène, Ijtti, Irchamene, Mijyene, Saïda, and Tizza (Figure 8.6).

The analysis of the results of the preliminary survey performed with a key informant of the commune revealed that at the level of the rural district of Boudinar, the social characteristics of the population by gender are almost identical, as women and men of the municipality practise almost the same activities to meet their needs. The population of Boudinar is decreasing as a result of migration. According to the results of the general censuses of the years 1994 and 2004, the population has slightly decreased during this decade with a growth rate of 0.7 per cent per year.

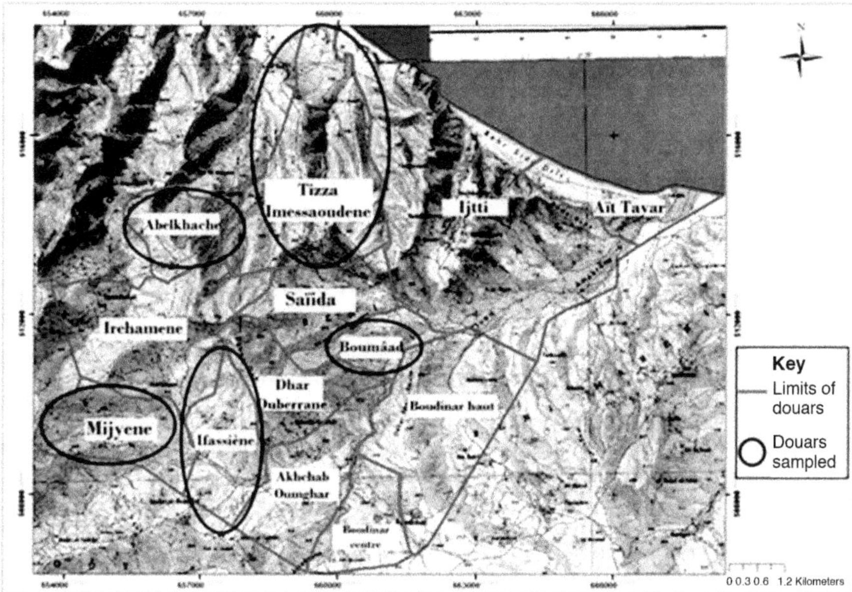

Figure 8.6 Boundaries of douars within the commune and the sampled douars

The level of education is relatively low especially for females.

The primary level schooling remains the most attended by men and women, with 48 per cent and 23 per cent, respectively. The number of children between six and 11 years old is about 1042 for males and 897 for females. The number of children belonging to the same age group and who are registered at the primary school is about 532 for male pupils (50 per cent) and 455 (50 per cent) for female pupils from 6 to 11 years old. On the other hand, 0.75 per cent of male children and 2.67 per cent of female children of school age could have gone to school, but did however, not join the school.

In passing from the primary level to the secondary one of schooling and in the university level, we notice a net decrease of the number of enrolments. This can be explained on one hand by the strong traditions of the commune preventing girls from attending school located outside the village and by the poverty on the other hand. According to the General Census of 2004 (HCP 2004), the number of male children attending high school is 2052, while for girls it is only 1795.

It is thus obvious that in Boudinar commune, there is a gap in the schooling of boys and girls as there are more possibilities for boys than for girls to attend school.

Although the majority of villages have primary schools, some male and female pupils interrupt their studies for various reasons. Arriving at the secondary level, the girls find a difficulty continuing their studies since the only available middle college in the commune is situated in Boudinar Centre, very far from the other villages. The average distance to reach the secondary school varies between 3km and 7km, depending on the villages. The pupils belonging to the category of villages (Ikhchamen, Mijyene) have to travel a long distance (7km) if they want to attend secondary education. In general, the main constraints facing children in the rural commune of Boudinar to continue their schooling are found to be the lack of roads and convenient transportation means.

Women are the less advantaged group of population

The statistical data reveals that in Boudinar there are several imbalances between men and women, always to the disadvantage of women. Men and women who are in their productive years (whose ages lie between 15 and 59) make up only 25 per cent of the Boudinar population. This means that approximately three quarters (75 per cent) of the total population in the commune are inactive. Among the ones who are active 50 per cent of males are considered to be working and only 4 per cent of females. Another imbalance is revealed as well in the proportion of single women, as it is found to exceed that of mens. While men's proportion of widows and divorced constitutes 0.6 per cent, women's proportion is 6.6 per cent.

Still another imbalance is in the realm of education, as more than 60 per cent of women are illiterate. Many girls either do not attend school because of traditions or if they attend, have to drop out from school to help in the housework and water provision for their homes. Very few women continue their education to high school or college, because of the lack of training institutions in the villages and also the lack of transportation and roads. This situation is exacerbated by the lack of opportunities in any income generation projects for women in particular. In addition there many physically handicapped women.

The physical handicaps coupled with the high illiteracy may further weaken the situation of women and limit their potential for active interaction in their communities. In a different vein however, the data from the health department reveals that 98 per cent of the pregnant women benefit from pre-natal medical services, and 83 per cent of women benefit from it after childbirth. These rates are considered satisfactory and prove that the majority of the population is conscious of the importance of pregnant women's health. Also, the majority of families prefer that pregnant women give birth in a medically controlled place. Indeed, the rate of infants born in the medically controlled places is equal to 71 per cent.

Methodology

Description of study participants

The results presented here draw on the accounts of over two hundred and fifty (250) male and female inhabitants, who participated in this study (100 in the quantitative survey and approximately 120 in the focus group discussions and workshops and 34 case studies). The participants in the quantitative survey and focus group discussions were drawn from the same portion of the local population according to the following criteria, that they:

1 are male and female
2 live in the area
3 live in a household with several years engagement in agriculture, and
4 aged 18 and over.

The households that meet these criteria are particularly vulnerable to climate variability and long-term change because with small land holdings there is limited scope for diversifying crops and their highly variable water resources expose them to uncertainty in their on-farm production. Such farmers have little margin of error in spending on high farming costs, which competes with money for purchases of food and non-food items. A broad age range of respondents and a requirement to have had several years of experience in farming were used to ensure that those with longer memories and substantial experience would contribute their experiences to the study.

We adopted gendered qualitative and quantitative approaches in this research.. Hence it adopts gender analysis which means that it studies the different roles of women and men in order to understand what they do, what resources they have, and what their needs and priorities are (FAO 2010). Stated differently this study uses gender analysis to document and interpret how men and women farmers perceive and respond to climate variability to ensure their livelihoods and food security. This report focuses on the gender roles and relations of husbands and wives, the primary decision-makers in the study households, while noting that gender roles are not static and are shaped over time by other differences such as age and class.

The research period lasted for two years: from January 2010 till December 2011 and was divided into three phases. During this period both qualitative and quantitative methods were used. The first phase comprised the qualitative type of research in which various participatory tools were used. Once the first part was over, the second part which entailed the quantitative type of research started, which was followed by the qualitative approach, during which the case study approach was used.

Methodological approaches

The use of multiple methodological approaches (sociological analysis of qualitative and quantitative data as well as climate analysis) helped to triangulate information thus ensuring that the farmers' reporting was substantiated and robust. The results obtained through one tool were then validated with a group of men and women as well as with key informant interviews.

Establishing direct causal links presented serious methodological challenges. While in some cases it is possible to link climate variability to changes in livelihoods, it is impossible to fully understand these dynamics without an understanding of the broader context, which requires a multi-disciplinary approach. By maintaining the focus on the farmers' perspective throughout all data collection activities, the results present a multi-faceted account of their point of view. During the research, the farmers were not directed toward a discussion of climate variability, but rather care was taken to place the research questions in the broader context of the farmers' lives and livelihoods. Thus a series of questions about the weather and natural resources were used to establish that farmers have seen changes in the weather and experienced severe droughts. From there the questions shifted to a focus on the impacts of the climatic changes.

Secondary data analysis (various censuses and documentation)

The first stage of this research work consisted of a preliminary survey on the socioeconomic context of Boudinar and on the state of water resources as well as an analysis of existing secondary data found in the documents of studies undertaken before in the study area. During this phase, the indispensable data essential for the study and for its organization was collected.

Qualitative focus group

Qualitative focus group discussions presented opportunities for men and women to express their views separately and together, which allowed cross-referencing and verification. All workshops and focus groups were sex disaggregated, they were either for men only and or women only. The attendees varied in age, social status and level of education, thus there were young and older men or women. Female attendees, however, were mostly either young girls or widows, but seldom married. This is because married women seldom participate in public meetings. They were however not excluded. Among the group of attendees in this study, married women constituted around 15 to 20 per cent of the participants.

As for the level of education among the attendees, most men or women were not educated, in particular the elder ones. Among the younger attendees there were some young persons who had attended school for some years (two to five years). In general in Boudinar, women were found to be more illiterate than men as pointed out before.

To collect the information during the workshops using a qualitative approach, several participatory tools were applied, namely: a social domain analysis, village social map, timeline, village resource map, daily activity clocks, seasonal calendars and farming system diagrams. These tools were used to facilitate focus group discussions with separate groups of men and women. Some of the tools were used to meet more than one research objective. Prior to the collection of data by the qualitative approach, all the 13 villages of the Boudinar commune were categorized into five main groups, depending on the availability of water sources and the number of their inhabitants. In each of those five categories of villages, focus groups of men and women were formed and we conversed separately with them. A douar is even smaller than a village, but everything is relative because it all depends on the definition of a village, district, and sub-district in every country.

Social domain analysis

Social domain analysis helped in the identification of learning opportunities that are structural, communicational, temporal or adaptive. It was used to examine how people view actors and relations between them using words and characteristics that participants themselves chose and defined. The procedure consisted of organizing several workshops in the five douars; to which most actors were invited. During the meetings, all the key stakeholders were identified, meaning those who may affect or be affected by the water scarcity in the area.

Village social map

During the workshops, participants were asked to show the location of all households on flip chart paper. Once all the households were illustrated, a group discussion followed on what constituted wealth and well-being until agreement was reached on the main criteria. The criteria agreed upon were: 1. the importance of livestock in the householdsn, and 2. the magnitude and usefulness of the agricultural surface owned by the households. Once these were decided, each household was assessed with these criteria. At the end, Socio-economic and Gender Analysis[4] (SEAGA) questions were used, to probe further about other household characteristics and differences, and population trends.

Timeline

The 'timeline' was used in conjunction with the 'problem analysis chart'. The purpose of a timeline is to help identify the events that have created a certain problem or situation. It explores people's views and knowledge about how a problem or a situation has evolved over time, and changes that have occurred in the process. In this case we used the merging of these two tools to help document the farm activities, food security and major livelihood opportunities, including migration, during the same decades as the seasonal calendar in order to overlay the climate conditions onto the livelihood activities and see how they relate to each other. In each focus group the participants were invited to discuss the areas of change in the village from their perspectives as well as focus on the core problem that constitutes a threat to the population's life. The results here were found to be identical between men and women as they both revolved about water and agriculture. For this merge, a table of 'before-and-after' was created with eight columns that describe the areas of change:

- (Column 1), the changes from what used to be
- (Column 2) to the present situation
- (Column 3) for each area, how important these changes are
- (Column 4), and then the causes or reasons
- (Column 5) behind each change, also the consequences of each change
- (Column 6), the techniques traditionally used to face each change
- (Column 7) and at the end any proposed solution
- (Column 8) to stop the problem.

Participants identified first all the areas of change in the village. Then, votes were taken to determine two priority problems. For each problem we tried to fill all the columns of the table. During the discussions of the causes, the events that had preceded these causes were identified in chronological order, and the groups discussed how those finally led to the current situation.

Village resource map

This tool was used to help identify the physical as well as social context of the community in order to characterize the local climate conditions and risks. In each category of villages, focus groups of men and women were formed and we conversed separately with them. Here the starting point was created by drawing our point of reference on the map, which was usually the school or the mosque where our meeting was taking place in. Then we asked the participants to draw other facilities and resources that they also considered important in the village. The participants were encouraged to work on their own with minimal interventions of the facilitators, which allowed free discussion among villagers. The facilitators only interfered when they needed some clarification, as for example if the water source was a private property as opposed to a common one, or to stir the discussion forward.

The water map was drawn and all the wells and springs inside the commune were identified and mapped for each village representing one of the five groups. The distance from settlements to the main source was estimated and the workload and time spent to collect water was roughly averaged for the whole groups of villages. We also assessed the water quality and availability. Finally, we used Socio-economic and Gender Analysis (SEAGA) questions to deepen the reflection on the social and economic uses of water, and the distribution of the roles and responsibilities of both male and female users in the community.

Seasonal calendar

This tool was used to describe the farmers' perceptions of seasonal climate parameters such as typical rainfall amounts and timing as well as average seasonal temperatures over the last years and until the present time. For this tool, male and female participants were met separately in focus groups, and were asked to discuss the activities they do throughout the entire year. In separated focus groups of men and women, we explained to the participants that we want to learn about what they do throughout the year. On a large paper, we began by drawing a line at the bottom and a parallel line at the top. This line was divided into 12 compartments each containing one month of the year.

We began by asking questions on the agricultural work, then on the waged work, then on the availability of the water and finally on the availability of time. Women were requested to add their reproductive work. For each treated theme, participants had to give the work one of three marks: one – weak intensity, two –average intensity or three – high intensity. Participants were invited to discuss the activities undertaken in each month and votes were taken based on the majority of the opinions to see where the mark should be placed. After the calendars were accomplished, participants were invited to discuss the influence and the availability of the water on the availability of time for the various tasks for women as well as for men.

Daily activity clocks

This tool was chosen to narrow the focus from the seasonal perspective to the daily one to look closer at the division of labor between men and women in their daily lives. With the focus groups of women and men separately and after having explained to the participants the purpose of the tool we asked them to tell us all their activities during a regular working day. We started by the activities of the previous day, we noted them all on a piece of paper then we plotted them on the circular pie chart (to look like a clock). We noted in the same portion of time the activities realized simultaneously. When the clocks were ready, we asked questions on the activities which appeared there. We assessed thoroughly two clocks for every sex: the clock of the activities during both of the seasons; Winter-spring and summer-autumn. Finally, we compared the clocks of both seasons. In order to initiate the discussion about gender concerns, participants were occasionally asked if they knew if the other sex, men or women are involved in this activity and why they do or they do not engage in it.

Farming systems diagnosis

While the tool of the daily activity clocks looked at the distribution of the daily activities of both men and women in a collective manner, the farming systems diagnosis allowed for a more accurate individual one, namely at the level of the household. It also relied more on merging with the observation tool. From each of the five village categories, a household was chosen and the research team spent a whole day and stayed overnight in order to observe the entire daily activities as they are divided among the family members. The use of this tool began with a visit to a farm, and then explaining to the farmers that we were interested to know all the various agricultural and extra-agricultural activities at the time of the visit as well as during other seasons of the year. During our stay we assessed the various water sources of the household, for example, if the household had a private water supply point and or a pond of water collection, thus ensuring that we had a complete picture of the persons in charge of the activity and the time spent on the practice of this activity. At the end of the visit, a meeting was arranged with all the family members, in order to recall and discuss what was seen.

Observations were made and recorded throughout the entire research period. The work was done during the stays in the communities: either during the overnight stay in the villages, or during the field research in the farms or even during the walks in the villages or while communicating informally with local people. Observations were noted also during interviews or after the field visits.

Case studies

This tool was used to focus more closely on the contemporary, real-life context, meaning to investigate more closely in a qualitative manner the detailed living conditions of men and women in Boudinar. The use of the tool aimed at exploring: a) if the interviewee noticed any change in the rainfall pattern over these last few decades and b) how and in what ways these changes affected them, their family members as well as their community? c) In case of crop losses, how they ensure food security at household level? d) What happens in cases of migration: when, by whom and to where? Who decides and what happens to the agricultural land?

The questions revolved around the distribution and division of the reproductive, productive and community management roles of the male and female interviewees amidst all these changing circumstances.

The quantitative approach was adopted in the second research phase and was of a non-participatory nature. Two prime tools were applied here, namely: the review of secondary material and questionnaires.

Review of secondary material

Secondary information was reviewed in order to validate the results obtained during the participatory workshops. Secondary material was reviewed using various sources of information such as climate analysis, the census of population and habitat (2004), Census of agriculture (1996), Official censuses for inhabitants and agriculture, existing monograph of the commune, past studies carried out in the area etc (The secondary data was obtained from documents listed in the references list). The climate analysis was useful as it provided climate indicators in the region to characterize the climate variability and trends of the past decades which actually exceeded the farmers' descriptions. Thus the review of the material was useful as it provided a broad overview on the commune. Most of the data obtained from the secondary material were discussed at the village level, for example types of houses, socio-demographic data, etc.

Questionnaires

Questionnaires were used in order to: a) verify the findings from the workshops, b) to quantify aspects that were identified and considered crucial for completing the information collected earlier, as well as c) to fill several gaps that appeared from the preliminary analysis of the qualitative information. Thus the questionnaires were used to collect quantitative data and deepen the analysis. In total 15 questionnaires were administered to a purposively selected sample using specific criteria drawn on a number of households by cluster, not only in the villages where the workshops were held but also in the other villages belonging to the same cluster. Three selection criteria were set to choose the households from the five village categories, namely: a 'fairly well-to-do' household, a medium one and a poor household. The exact selection of each category was done randomly. In most of the cases, the village head was the one to recommend the households from among each category. We administered the questionnaires to 100 individual respondents. Views from members of the same household were collected to control for variability between households. In addition, female heads of households also completed the questionnaire. Almost all of the respondents were from households that owned rainfed land, while many of the respondents owned some irrigated land as well as rainfed land.

Results and discussion

This section presents the main research findings, and then discusses how they compare with other findings in the literature.

The gender division of labor is unequal throughout the year and is strongly related to water availability.

The analysis of timetables and seasonal calendars of men and women reveals that the population of Boudinar performs numerous activities all of which are necessary for the sustenance of their livelihoods. Indeed, the productive and reproductive activities, which are unequally exercised by men and women during all year, have varied rhythms according to the seasons.

Reproductive roles

The domestic work is essential within each household. Being an exclusively female responsibility, it constitutes an inevitable load for these women and it has a high intensity all year round (Table 8.3).

Provision of water remains an activity that is almost exclusively done by women and girls. This activity becomes more intense and more exhaustive during the dry period from June until October. In addition to taking care of their homes throughout the whole year, women also perform productive roles.

Table 8.3 Reproductive activities profile in the commune of Boudinar

Reproductive activities	*Women*	*Men*
Fuel/wood collection	++	+
Water collection	++	+
The preparation of food Domestic tasks (cooking, washing, etc.)	++	–
Children and elderly care	++	–
	++	–
Women and men contributions to the reproductive activities	10+	2+

Key: Degree of intensity/importance for women and men: ++ intensive, + less intensive, – do not perform the activity

Productive roles

In general, the agricultural work, in which both men and women participate, is performed throughout all the year, but becomes intensified during the six months from October until March. Activities related to animal production are also practised throughout the year and require a considerable amount of time especially from women. Waged work, which some men and women resort to, is practised all year for some or according to the intensity of the agricultural work for the others. Indeed, it is the members of the families who do not possess private agricultural lands that turn inevitably to waged work either in farms in the same village or even in the closest cities. On the other hand, during the period when the agricultural work becomes intensified, some men, women and girls may turn to work in larger farms in their neighborhood to increase their income. For the women, this is true especially during the period from November til February when the work of weeding and sowing increases, as well as during the period from May til July, coinciding with the harvest of cereal. For some men, once the agricultural work decreases, they try to work as workers in the commune or in the closest cities (El Hoceima, Nador) in order to increase their income.

The preparation of lands which requires plowing and sowing is considered a men's work, and is performed in autumn and in winter. Hence men's seasonal calendar is found to be loaded during these periods. In autumn, the picking of olives takes place and is generally undertaken by women. During spring and summer, female farmers are occupied in the picking and the transport of the harvest. While the workload of men decreases during the summer period, for women their reproductive work remains constant throughout the whole year. Hence they are prevented from enjoying any leisure time, unlike men who can do so then. The findings of the research have recurrently revealed that the time duration and the total workload differ drastically among men and women. The analysis conducted on the workload of both men and women throughout the year shows that women work more than men throughout all seasons. The example of Boumaad village in Table 8.4 illustrates the main activities performed by men and women in each period of the year, and highlights the activities that are solely by men or by women.

Looking at the previous table we can assume for the gender division of labor in relevance to water related activities that:

- domestic work is entirely women's responsibility and it increases when there is a shortage of water.
- during shortages of water, women's free time decreases drastically while men's availability of time increases.
- during times of improved water availability, agricultural work increases for both male and female workers. For women this means a double burden, an increase in agricultural labor as well as domestic tasks.
- during times of water scarcity, women's opportunities for waged work vanish completely, while male laborers retain a low intensity, waged work opportunity.

Seasonal calendar showing main activities of the women and the men in Iffasiène

Table 8.4 Distribution of women's and men's workload around the year

	Winter (December, January, February)	*Spring (March, April, May)*	*Summer (June, July, August)*	*Autumn (September, October, November)*
Women	– repair of fields for the sowing and provision of fertilizer; – collection of wood for fire(light) and cooking; – help in the sowing of the truck farming and some cereal	– picking of the truck farming: legumes, charm – weeding of fields; – harvest of cereal (wheat, barley)	– harvest of cereal (wheat, barley); – picking of beans (charms) and peas, lucerne, potatoes; – maintenance of houses	– preparation of fields for the sowing and the contribution of the fertilizer; – picking of olives and their cleaning
Men	– sowing of potatoes, onions, leguminous, peas, Lucerne and cereal; – plowing	– picking of the truck farming: the legumes – harvest	– harvest; – waged work; – beating and tossing	– sowing of potatoes, legumes, peas

Table 8.5 Seasonal calendar of women in the village of Iffasiène

		J	*F*	*M*	*A*	*M*	*J*	*J*	*A*	*S*	*O*	*N*	*D*
Availability of the water	M	1											
	W	2					1				2		
Agricultural work	M	2					1						2
	W	3							2		3		
Waged work	M	1				2				1			
	W	–			1				–				
Availability of time	M	1					2						1
	W	1											

Key: 1= Weak intensity; 2 = Average intensity; 3= High intensity.

For both men and women water is rare in the village, especially for the inhabitants who are far from the main spring. For men, agricultural work becomes intensified moderately from December til May. The waged work is performed throughout the year, but becomes intensive during the period from June to November. It is the period during which the agricultural work decreases. In general men are found to possess more time in summer and in autumn. This availability of time depends on the agricultural work when it becomes less intense. As a result, men can benefit from this spare time to perform other tasks. With regards to women, a striking observation can be made, namely that for them the agricultural work remains intense throughout the whole year. In a similar vein women are found to participate in all the agricultural activities with certain tasks being specifically confined to them, such as weeding. During the period from April to July, certain women can work as farm laborers with their neighbors, most likely in their own village. Women's spare time is very limited all year because of their heavy workload.

Iffassiène village (Table 8.5) shows how women's workload exceeds by far that of men. This is because the various activities are not shared in the same way between men and women. Women are sometimes found to perform agricultural activities which men do not contribute to. In addition to that women perform the reproductive activities such as the collection of the water and the collection of the wood, which they perform on their own.

The Iffassiène case highlights three main findings:

- with the low availability of water, women's work becomes more intensive as they are engaged in both reproductive and productive roles.
- when men's work share doubles in the agricultural work, women's share triples.
- when men's income opportunities double, women's share decreases. Thus men's share of waged work, which provides income generating opportunities, or leisure time, doubles that of women.

The unequal distribution in men's and women's daily routine in the gender roles: the productive and reproductive, denies women access to time as an important resource.

The analysis of the daily routine diagrams of men and women in the five sampled villages demonstrated several issues:

The distribution of activities practised by men and women is similar across villages and through the four seasons. Women's daily routine, however, is loaded compared with the timetable of men throughout all the seasons, as women's days are packed fully with activities from morning to evening. In general women begin their days with both reproductive and productive tasks, in particular the care of children and attending to animals, water collection, domestic work and collection of wood. These activities are carried out prior to going out to agricultural fields to help men in farm work. During afternoons, women have to finish all domestic work before going out once again to work in the fields. Before nightfall, some women who do not have enough water in their homes leave for the second time for fetching water. The frequency of water collection increases during dry seasons, when water becomes scarcer and the demand for it increases.

Men, on the other hand, work from morning till the end of the afternoon in productive activities related to agriculture. In the evening, when men can have leisure time in their homes or in coffee shops, women are occupied with various domestic activities, such as preparation of meals. Men are thus only involved in agricultural work and can benefit from spare time which they either use to rest at home or meet with other men in the village. This also allows for social networking and information exchange which women do not have.

Community management activities, as will be discussed in detail later, take place outside Boudinar Centre and address only men. In the village of Boudinar Centre, several activities with relevance to water provide a good opportunity for people to participate collectively. Findings, however, reveal that these activities are only attended by men. This is because women are caught up in their reproductive duties in their homes. Thus they try to finish their chores early at the beginning of the day, so as to have some spare time in the evenings, in order to practice some leisure activities such as sewing and or education, etc.

Daily time

The distribution of the daily tasks performed by men and women in the rural commune of Boudinar reveals that women tend to have longer and more loaded days than those of men.

Indeed, women are found to contribute to the agricultural work which begins very early in the morning, after they had collected the water, and after they have taken care of their children and their animals. Likewise during the evenings, women and girls take care of the animals, collect water and wood as well.

The diagram (Figure 8.7) illustrates the daily routines of both men and women in the village of Abdelkache.

In conclusion, it can be maintained that women contribute to the agricultural activities and animal breeding besides their multiple reproductive works both within and outside their homes. On the other hand because men are engaged only in the seasonal agricultural works, their timetables can be more relaxed when the agricultural work decreases. As highlighted in the previous section, it is again verified that women are the only ones deprived of their spare time throughout the year.

Inequality in the distribution of community management roles denies women access to several resources such as information, technology and decision-making.

Community management roles

Community management roles entail being active and willing to serve at the community level, attendance at and organization of social and cultural events. According to the findings, there are four which perform a few services for the local population as follows:

- The Fouss-G-Fouss Association for culture, the environment and social development, is located in Boudinar Centre. Its interest is mostly focused in solid waste collecting. This association has some women members. It has already led

Daily routine activities in Abelkache village

Figures below correspond to the daily routine activities of men and women during the season winter-spring.

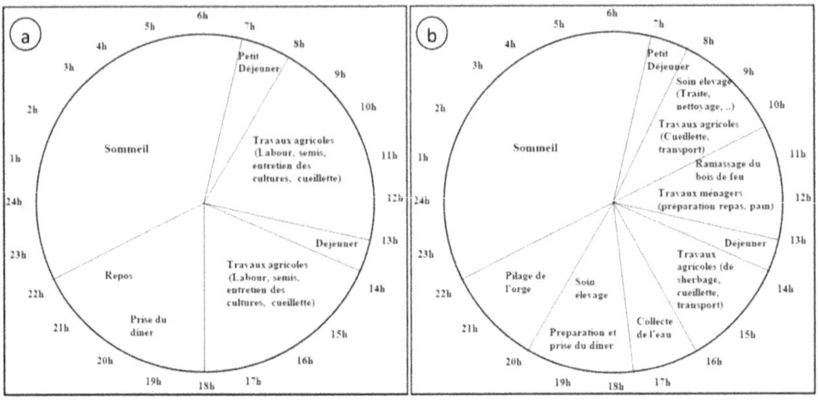

Figure 8.7 Men's daily activity routine in winter (a) and women's daily activity clock in winter (b)

During the season winter–spring, the women wake up generally at 7 am and take one hour for the breakfast and to feed the cattle. During the morning, the women take care of their cattle, clean the stable and feed the animals. Later, they have to go to the fields to help the men in the picking and the transport of potatoes, peas and legumes.

When the agricultural field of the family is not far, the women return home to prepare the lunch, the bread and do the domestic work. It is during the morning that women do the collection of the firewood.

In the afternoon, women collect the water as well as care for the livestock. Some girls have to bring out the cattle for the pasture and go to the forest of El Tletha. Dinner takes place around 8 pm; the women don't have enough time to rest because before they go to sleep at about 10 pm, they complete some domestic work or take care of the pounding of the barley.

The daily routine of the men in the same season is much more relaxed. While men wake up and sleep at the same time as women, they have more time to rest. Moreover, men's activities are not as diverse as those of women. Their activities are found to be limited to the agricultural fields, where women are also working simultaneously. During this season, men spend their days plowing, sowing and performing maintenance of the cultures or picking, according to the month of the year. They generally finish their work at about 6 pm and take advantage of the remaining time of the day to rest.

various activities. However, it is confined to very few local inhabitants of Boudinar Centre, who meet there.

- The Tiniba Association located in Boudinar Centre is a female association which was created with the aim of supporting the capacity building of women by teaching them some skills which they could use to generate their own income. Most of its members are women and young girls. It led some modest activities such as combating illiteracy, provision of training on sewing, embroidery and domestic decoration. This association also serves only the inhabitants of Boudinar Centre or neighboring villages, who are able to go to it occasionally. However, this association becomes more and more inactive because of the lack of finances.
- The TAFADNA Association for the environment and development is located in Tizza and aimed at the social, economic and cultural development of the region, and to the preservation of the environment and the encouragement of solidarity actions. It thus manages the provision of potable water for homes. Even though it deals with a problem which is mostly of interest to women as they are the ones responsible for water collection, women are not visible in its activities. So far it has initiated projects in the domains of saving the drinking water, and the eradication of illiteracy. This association is still active but benefits only inhabitants of the village Tizza. Both the Associations of Tafadna and Fouss-G-Fouss are putting together a community-based project for Boumaad Tizza to reduce the impacts of climate change.
- An association was created at Boumâad, but it is not active because of a lack of finance. This association is composed solely of men, and its interest is in agricultural development and provision of water for irrigation. Indeed, it intended to initiate some works for reorganization of a waterworks which was built by the local population of the village, but as yet no activity has been implemented.
- Finally, there is the Association of 'Touiza', which is known for its strong sense of solidarity. It implements community works, such as the maintenance of infrastructures, and encourage mutual aid in agriculture among the locals. The Association has set up an animal production project that failed due to lack of financial and material resources to carry out monitoring and training of women in charge of the project. Due to worsening living conditions, its members have migrated, thus weakening the strong social bond of the association as well as the livelihood of the villages concerned.

Throughout the five associations, the presence of women in the decision making process is not visible and men play a dominant role. Female community members' participation is indirect, they express their points of view through their husbands or through other male family members.

Recently the Moroccan government has issued a decree necessitating the presence of women in the boards of the communes and hence there have been two women that have been elected as members of the board of the Boudinar commune.

The commune of Boudinar also supports the development of rural women through the creation of income generating activities. For example a training center for young women teaches literacy classes, sewing and baking.

From the findings it becomes obvious that within the realm of community management, women are seen as beneficiaries instead of active change agents. Men and women have different levels of access to institutional support and women have limited access to information relevant to farming in comparison to men. The gender biases created through the societal roles expected for females and males are imported to within the institutions, which in turn often reproduce assumptions as argued by Gurung *et al.* (2006), that it is men who are the farmers. As a result, not only are women's voices not heard, but new agricultural information, technologies – including the replacement of plant types and animal breeds with new varieties intended for higher drought or heat tolerance – are found to be rarely available to women farmers (Lambrou and Piana 2006).

Evidence of climate change

The region of Boudinar suffers from several problems related to climate change. These include intensification of cold and freezing which destroy agriculture, specifically the potato. Flooding destroys the water canals and leads to reduced soil fertility, isolation of the whole commune, destruction of houses, power cuts, lack of potable water, loss of resources and incurs diseases. Excessive winds result in destruction of crops and foliage plants and fruit trees. The inhabitants of this region are threatened by damage to all these phenomena, such as: the destruction of houses due to heavy rain and flooding. The extreme weather means they have to make adjustments and repairs each year (replenishing lime and building roofs).

The bad weather results in plant disease and loss of crops and hence increased use of pesticides.

Several other constraints heighten the problem such as: lack of a training institution, lack of means of transport and illiteracy, which results in an inability to obtain information regarding agriculture. During scarcity of water, the problem becomes even fiercer as water prices rise.

Certain other factors, however, positively influence this change variability, namely: the available supply of drinking water in some douars through tab pipelines, the financial support provided through families living abroad who send remittances to their relatives, and the opportunity to work in the construction of buildings, which remains uninfluenced by the effects of climate change.

Gendered forms of differential vulnerability

Men and women experience changes in climate variability differently. This section discusses five different forms of the variability.

Gendered dimensions of water use and management and climate change

As numerously noted in the gender and environment literature, and confirmed by our findings, women and girls generally assume primary responsibility for collecting water for drinking, cooking, washing, hygiene and raising small livestock, while men use water for irrigation or livestock farming (Fisher 2006; Khosla and Pearl 2003). These distinct roles mean that women and men often have different needs and priorities in terms of water use.

But while this knowledge isn't new, it does take on a pressing significance in the context of climate change. Climate change leads to several phenomena. This has a particularly harsh effect on women and girls because of their distinct roles in relation to water use and their specific vulnerabilities in the context of disasters. In Boudinar, during the previous years of long drought, many wells were dry and others became salty and numerous sources of water supply were abandoned. As a result the time absorbed by water collection increased as women spent more time fetching water from remote sources. Irrigation water as well was less available. Women then were the ones collecting water on their donkeys, and the laundry and all the cleaning of farm produce were done on the river.

At present, water is either obtained through tap water or rains and is stored in large tanks some of which are built inside houses. Collecting water is only in times of cutbacks. Thus the time allocated for water collection has changed; it has decreased due to the increased use of home tanks and the availability of water in winter. As there was more rain over the last four years, water became available in the wells, and thus the time for waiting to collect water became relatively short. In the summer however, the time allocated to water collection has increased, because of water shortages and scarcity of spring water. In summer girls may go to the source four times a day and carry, using animals, up to 280 liters per day over a distance that might take an hour to walk.

Over the last four years, Boudinar region witnessed the increasing frequency and intensity of floods and deteriorated water quality. The heavy rainfalls over the last years and the more frequent floods resulting from climate change have nevertheless increased women's workloads, as they have to devote more time to collecting water, which increased because of the difficulties produced by the flooding, especially during rainy periods when roads are impassable. Moreover women have to spend more time on cleaning and maintaining their houses after flooding. This time could be rather spent in school or earning additional income.

Gendered dimension of agriculture and climate change

As maintained by FAO (2003), rural women and men play complementary roles in guaranteeing food security; nevertheless, women tend to play a greater role in natural resource management and ensuring nutrition. In Boudinar as well, as findings have revealed, women often grow, process, manage food and other natural resources, and are responsible for raising small livestock, managing vegetable gardens and collecting fuel and water. Men, by contrast, are generally responsible for cash cropping and larger livestock. Thus because of women's involvement in agriculture, they are most adversely affected by the impacts of climate change, because the responsibility for adaptation is likely to fall on their shoulders – including finding alternative ways to feed their family (CIDA 2010).

In Boudinar earlier climate variabilities of droughts have resulted in changes in plant production and farmers were forced to increase crop rotations and thus dedicate more time to the agricultural work. There is also a remarkable change in the quantities of the plans that have increased through the planting of fruit trees in the Green Morocco plan and abandonment of cereals. Farmers had to diversify the products, and focus on products that support dry weather. Accordingly many are found to drop out of irrigated land and orient themselves more toward the rainfed land, which requires less care. Though there is a momentary decrease in resources and increased expenses, incomes are relatively stable and families are able to cover costs and needs from the revenues of the sale of replacement products. Nevertheless, agriculture in the region can't be considered a safe resource, as the floods of the Wadi Amkrane and plant diseases threaten the food crops. There is also a noted decrease in the quantities and prices due to crop losses, increased expenses and decreased purchasing power of the community. Farmers are forced to sell cheap products on the market because they cannot afford to store them. A farming household has been calculated to pay at least 2,000 dh[2] per year for pesticide treatment in order to fight against plant diseases and the loss of the crops.

Farmers' adaptations over the last five years, with the availability of water in wells and springs, have been through pumping water from the river and wells, and the use of pipes to bring water to the fields. Likewise farmers resorted to the system of channels directing water to different irrigated fields (seguias), which, however, are often damaged by floods and river flooding.

As for the gender division of labor, women are found to be the primary gender responsible for harvesting. Women and girls, who previously used to participate with men in the harvesting of agricultural products, became the prime ones responsible for the collection of olives. Thus the collection of olives has replaced the harvest as a source of income for women workers. When the harvest is not good, there is less work in agriculture and less access to resources for both sexes and less fodder for livestock. Women trying to find a substitute for feeding the family resort to the provision of fodder for cattle and livestock, the main source of income for women. While some men remain engaged in the fields, other men have left agriculture altogether to work in building construction.

Gendered dimension of health and climate change

Confirmed by the literature, the findings reveal that in Boudinar, women and girls are generally expected to care for the sick, particularly in times of disaster and environmental stress (WEDO 2007). This limits the time they have available for income generation which, when coupled with the rising medical costs associated with family illness, heightens levels of poverty. Other likely health consequences of climate change include higher rates of malnutrition due to food shortages, increases in heat-related mortality and morbidity, and increased respiratory disease where air pollution worsens. Children under five are the main victims of sanitation-related illnesses, and – along with the elderly – are most affected by heat stress (Bartlett 2008). In addition, being faced with the burden of caring for dependents while being obliged to travel further for water or firewood makes women and girls prone to stress-related illnesses and exhaustion (Voluntary Services Overseas 2006; CIDA, 2002). In Boudinar, the residents when they are sick use the public hospital because it is cheaper. Women and girls, who face barriers to accessing healthcare services due to a lack of economic assets to pay for healthcare, as well as cultural restrictions on their mobility, may be constrained from travelling to seek healthcare. This also means that women, as the primary carers of the sick, have to spend all their time in the care of their sick family members. Hence they are less able to contribute to community-level decision-making processes on climate change or disaster risk reduction.

The elderly in Boudinar as well, as revealed in numerous cases studies, are at highest risk from climate change-related health impacts like heat stress and malnutrition. Elderly women are likely to be particularly vulnerable, especially in times when resources are scant and social safety nets limited or non-existent. Elderly women may have heavy family and caring responsibilities which cause stress and fatigue while also preventing wider social and economic participation; and their incomes may be low because they can no longer take on paid work. They may also not understand their rights to access local healthcare services. Even when they are aware of these services, even nominal amounts for clinic visits and drugs may not be affordable. Access is further restricted for older women, primarily the handicapped ones living in rural areas, who are often unable to travel the long distances to the nearest health facility. Older men are particularly disadvantaged by their tendency to be less tied into social networks than women and therefore unable to seek assistance from within the community when they need it and hence become even more dependent on the female members in their households, adding to their confinements and chores.

In a similar vein, young girls and boys, who usually can carry up to 30l of water, two to three times a day, can suffer during period of flooding or major rainfall as the trails get rough and are cut and also in summer during periods of high heat. A decline in food security and livelihood opportunities, can as well cause considerable stress for men and boys, given the socially ascribed expectation that they will provide economically for the household. (Masika 2002).

Gendered impacts of climate change on wage labor

As noted above, women's access to economic resources in terms of income, time, education, work opportunities and knowledge is already often unequal. Also among Boudinar farm workers, a gender gap in earnings persists in informal wage employment and self-employment. Thus women, who comprise the majority of those working in the informal employment sector which is often worst hit by climate change-related disasters and other shocks (WEDO 2007), and who already have unequal access to resources, have to face with their diminishing capacity, the unexpected events/disasters or adapt to change.

During the recurrent periods of rain and flooding, the seguias (water channels) were destroyed and crops were lost. As there is less work as agricultural laborers, men abandon agriculture and go to work in the building trade as day laborers. Women, who were left in the villages, diverted their attention to livestock, raising chickens and rabbits for family needs and for market, and sale of livestock products (the milk, butter, whey, goat).

Due to the decrease in agricultural land and change in products from seasonal crops to olive trees, which require five years to be productive, women have to resort to feeding their cattle in the mountains, to which they have to go on a daily basis. In some cases the distances require three hours going and three hours returning. Thus the job opportunities reserved for women as sources of income have become scarce and are casual and low paid.

Gendered impacts in climate-related disasters and coping strategies

As revealed before, already during the 1980s in response to severe droughts, several male farmers started migrating and many remained away permanently. Over the years and in response to low farm income due to droughts or increased floods, farmers resorted to wage employment, and male farmers in particular to migration, as they realized that changes in the climate were making farming an unreliable source of income. The income from wage employment enabled them to access food through purchase. Today men migrate to cities for construction work during the non-agricultural season, yet this time not as a permanent migration.

In Boudinar there are various other problems that are coupled together in the market scene. There's an increase of unemployment, and many boys refuse to work on the land and prefer to engage in informal businesses. Second, aid from migrants abroad used to be an important source of income for the inhabitants of this region. This resource has now decreased, as migrants' social and economic conditions have changed. Thus two important sources of income have disappeared with no replacement whatsoever.

As mentioned earlier, because of lack of resources and increased expenses, inhabitants' interests diverted more and more to livestock raising. As a result, there are fewer resources for women now as animal products, including milk, butter and whey, which were used for sale, are used to meet the needs of the families. This in turn constitutes a great loss to the families with limited means.

Coping strategies in relation to food security

The amount of income from farm produce, which was the community's primary source of income, followed by wage labor, varied from year to year and was linked to climate variability. Income from farm produce was impacted, in order of severity by the weather, yield, prices and labor availability. This suggests that farmers' food security varies from year to year along with income. Not only is the production very much affected by climate change but the food security dimension is also likely to be affected.

In case of flooding, Boudinar commune is completely isolated and access to the souk (market) during rainfall becomes difficult. Thus the families use the stored products. Storing food is primarily a women's activity and hence they are the prime decision-makers on the type of food for consumption or storage. The food is stored in a traditional way and there are families who have a refrigerator for freezing meat and fish. Agricultural products are stored in plastic bags, such as potato, and meat in the refrigerator. In some cases, women use insecticides to prevent the stored products from rotting. In principle, households consume the stored products, for example the stock of wheat is used for a year and barley for six months and food is purchased from the souk on a weekly basis. Boudinar inhabitants always have some food storage to avoid problems caused by isolation.

The research findings reveal that gender inequalities can also be exacerbated in the aftermath of disasters. During the period of flooding, the activities of production and reproduction decrease, as no agriculture, no cleaning and no collection of water takes place as the river cannot be crossed. During the floods, however, women start instead to get involved in the evacuation of rainwater, repair of roofs and furniture, as well as care for the animals. After the floods, women begin the development work of the roofs and walls, clean the barns and evacuate water from the rooms to save their homes. Thus the entire household workload may increase substantially, and many girls might also be forced to drop out of school to help with these chores (Davis *et al.* 2005).

How results explain and expand our understanding of the problem

The gender-sensitive approach employed in this research revealed many dynamics of vulnerability and coping at household level that are key to overcoming changes in the climate. By speaking with men and women separately, their different perspectives and perceptions of risk emerged, as did their distinct areas of knowledge. The use of the gender-sensitive approach also served to reveal how changes in climate variability impact on dimensions of lives and well-being beyond simple direct impacts on water availability and crop production.

Further along this line, the findings of this research confirm previous gender findings as stated by FAO (2010), which indicated that the needs and priorities also differed between men and women in relation to climate change adaptation needs and priorities. Thus while female respondents in Boudinar during their discussions about the main problems affecting the community and their needs and priorities focused on the micro level, meaning their household's health and well-being; male respondents focused more on the macro scale. These findings about the needs corroborate as well with the findings of research on poor women of the Ganges River basin in India as stated by Mitchell *et al.* (2007: 16) as follows:

- a safe place to live: solid houses built with a high plinth level to reduce inundation and shelters required for people, animals and agricultural inputs/products;
- better access: to climate change information and related knowledge and skills, to services, such as doctors and veterinaries and to communications, through safer roads;
- other livelihood options: through knowledge and resources for crop diversification and adaptive agricultural practices.

Men's proposals were in line with how to develop agriculture and reduce migration of people.

Thus they emphasized the uncontrolled well drilling, which on the one hand causes a hazard to groundwater in the region, but on the other hand is replenished through rainfall in the same area. However, there was also some waste of water through runoff.

They requested the conservation of water resources, and carrying out of works to capture water and feed it into the water table. They suggested setting up large irrigation canals that allow an effective and efficient flow of rainwater and encourage farmers to use drip irrigation. They also requested the establishment of a system of water supply and power systems for the town of Boudinar that could withstand earthquakes and floods and techniques for collecting spring water and rain.

A male respondent in the Boudinar commune provided a proposal to sensitize policy makers on gender issues and then conduct a study about local needs and opportunities for women and forming women's organizations. It is only after that, that literacy, sewing and cooking classes could be offered. The respondent suggested as well the promotion of information on international negotiations on climate change through communication at local level by specialized agencies or government associations to be supervised by experts. The municipality, however, he maintained, would not be able to provide any human and material resources to support this task.

The fact that women focused their priorities on the well-being of their families helps us in the clarification of two points. One, as noted by the Action Aid/IDS report (Mitchell *et al.* 2007: 14):

> They [the women who took part in the research] might not be aware of all the possible adaptation strategies, of all the ways to overcome constraints to the ones they are using, but they certainly know their present situation best and have an urgent list of priorities to secure a livelihood in the face of the new challenges.

It is vital that these priorities are made visible and are used to inform policy decisions and programs on adaptation.

Two; and more significant, it re-affirms our previous points made, which state that in the present situation, women are the only ones responsible for the reproductive tasks and survival of their families. It also proves that women are limited in their access to resources. Hence, they are not the decision-makers of the lands, as they do not own the land title, and due to their limited time and knowledge, they do not participate at the decision-making level of the community. Accordingly their awareness is meager and limited to the micro level, to which they can only refer in terms of climate variability.

Future resilience

However, as rightfully asserted by FAO (2010) and which we need to endorse here, is that not all change should be seen as a threat. As the known patterns of agriculture undergo changes and the farmers seek to adapt, in this 'transitional adaptation phase' there may be emerging new empowerment opportunities for women. As new resilience patterns are emerging both at the institutional and household level, the gradual breakdown of barriers to women's empowerment is occurring. Because external threats demand new solutions, it becomes increasingly impossible to continue 'business as usual' in agrarian rural communities and hence opportunities emerge for women and men to participate in untried decision-making processes as well as to take on newly defined leadership roles. Gender relationships, especially among younger generations, are also in flux, which creates new spaces for social change as well as the emergence of new social patterns and opportunities.

In Boudinar during previous times, only men went to work abroad and due to cultural constraints left their wives and children back in the douars, in order to prevent their children and especially girls from exposure to foreign cultures. In present times, boys migrate mainly to the large Moroccan cities to seek work outside agriculture. As an impact of migration, women have become more and more responsible, and learned to invest more in agriculture and animal husbandry.

In a similar vein, age has a direct impact on gender roles: the older the woman, the greater her decision-making power.

At the level of the household, in terms of coping strategies, findings reveal that it is the grandmother and or the father who are the key people in the family who possess the knowledge and are referred to for the decision-making process.

Development and policy implications and recommendations

In this research the coping strategies identified are not equivalent to climate change adaptation, but by demonstrating men's and women's abilities to react to changes in the climate, the findings can inform our understanding of what is needed for longer-term climate change adaptation and the development implications. This is because they help illuminate how men and women do and do not cope with changes in their livelihoods and food security induced by climatic factors. Understanding how men and women have responded to climate variability in the past and how they respond now, including what external support assists them, is a crucial baseline for designing strategies for long-term adaptation.

Furthermore, the above leads us to pose a few additional questions about the way women and men farmers behave in conditions of climate change: how will men and women farmers fare? In particular, how will men's and women's roles in decision-making and their access to resources shape future livelihood activities? Furthermore, how might gender roles and relations change along with climate change? At the same time changes in access to resources at the household and individual levels due to changes in climate would also need to be addressed (FAO 2010). As women become increasingly responsible for the household income when the husband migrates, as mentioned earlier, will this signal the greater empowerment and agency of women as also maintained earlier? It would appear that as their social roles shift, women can increase their confidence in dealing with other daily sociocultural inequities they face. However, research in Australia for example (Alston 2007) suggests that this would generally occur if external institutions supported and made visible the increased power of women in agriculture through overt processes of information sharing and through training and development programs addressed specifically to them in a regional or national context. Without these external resources for women to become full-fledged farmers, their empowerment remains questionable.

The way forward and key recommendations for future research

Even though there are some excellent studies that trace the links between gender roles and climate change as presented, and because of the need for policies and programs to respond to the different roles that women and men play in the various circumstances, there remains a clear need for more in-depth research. In order to shape effective policy and practical interventions, research into other impacts of climate change as well as poor women's voices and experiences is needed as follows:

- Research is needed to highlight the effective mitigation strategies that are already in place which can provide models of best practice for communities in disaster-prone areas. For example, in La Masica, Honduras, there were no reported fatalities after Hurricane Mitch, because a disaster agency had provided gender-sensitive training and involved women and men equally in hazard management activities, and women took over control of the early warning system. This led to a quick evacuation when the hurricane struck. The above example demonstrates how a gender-sensitive strategy was the key to an effective response that saved the lives of both men and women (Bridge 2008).
- The United Nations Development Fund for Women (UNIFEM 2008) for example, has developed a set of *gender-sensitive early warning indicators* which reflect the projected effects of climate change on communities – particularly around depleting resources.

There is an urgent need to clearly identify the obstacles to women's participation in decision-making, and to find ways to address these constraints through supporting grassroots awareness-raising, confidence-building, advocacy and leadership training programs. It is by now widely accepted that failure to include women and especially girls in decision-making processes around climate change mitigation and adaptation at all levels exacerbates gender inequalities, and also undermines the effectiveness of climate change responses. This is because girls may be doubly excluded from decision-making processes and on the accounts of being both a child/youth and a female.

Future research questions could address the extent programs aimed at mitigating environmental impacts have included women. Or question the levels of female participation in decision-making on climate change quantitatively and qualitatively. The barriers to participation and what can be learnt from existing literature on promoting women's and youth's participation in decision-making could also be investigated (Bridge 2008).

Finally, more research based on gender analysis that recognises the heterogeneous nature of men and women farmers, and the variety of strategies which they develop and employ continuously as well as the type of gender-sensitive information and approaches needed to ensure farmers make appropriate forward looking decisions (FAO 2010), as it would enhance policy recommendations. Further work along these lines, that also incorporates the important dimensions of age and class (Ray-Bennett 2009), could increase the terminology for understanding the approaches needed in dealing with short and long-term climate variability and will increasingly provide more rigorous outcomes.

The differences documented here by gender in perception of climate change and water use, show the importance of consulting both men and women when seeking to understand how livelihoods may be related to climate shifts, both short and long-term. What the research zealously reveals is that gender does matter when assessing farmers' responses to climate variability.

Note

1 Abdellatif Khattabi, École Nationale Forestière d'Ingénieurs (ENFI), Salé.
2 Haut Commissariat aux Eaux et Forêts (HCEF), Chefchaouen.
3 Consultant, Cairo.
4 Socio-economic and Gender Analysis (SEAGA) is an approach developed by the Food and Agriculture Organization (FAO) in partnership with others to develop the capacity of development specialists and humanitarian officers to incorporate socioeconomic and gender analysis into development initiatives and rehabilitation interventions.

References

Anonymous (2010) 'Monographie Communale de 2010 de Boudinar Elaborée Conjointement avec la Commune Rurale', Ministère de l'Aménagement du Territoire de l'Eau et de l'Environnement (MATEE).

Anonymous (2001) 'La Situation Actuelle du Littoral Méditerranéen Oriental SDAULMO', Ministère de l'Aménagement du Territoire, Maroc.

Bartlett, S. (2008) 'Climate Change and Urban Children: Impacts and Implications for Adaptation in Low and Middle Income Countries', International Institute for Environment and Development (IIED) Human Settlements Discussion Paper – Climate Change 2, UK: IIED. Online at www.iied.org (accessed 2012).

BRIDGE (2008) 'Gender and Climate Change Mapping the Linkages: A Scoping Study on Knowledge and Gap'. BRIDGE Institute of Development Studies, University of Sussex, Brighton

Canadian International Development Association (CIDA) (2010) 'Gender Equality and Climate Change: Why consider gender equality when taking action on climate change?' Online at: www.acdi-cida.gc.ca/INET/IMAGES.NSF/vLUImages/Climate%20 change3/$file/Gender-2.pdf (accessed 2012).

Couprie, P. (1998) 'Essai de Cartographie des Zones Sensibles à l'Erosion dans le Bassin de Boudinar (Rif nord-oriental Marocain): Utilisation Combinée d'un SIG et de la Télédétection', Université Paul Valéry, Montpellier.

Davies, R. (2003) 'Network Perspectives in the Evaluation of Development Interventions: More Than A Metaphor', EDAIS Conference 24–25 November 2003, New Directions in Impact Assessment for Development: Methods and Practice.

Davis, I. *et al.* (2005) 'Tsunami, Gender, and Recovery – Special Issue for International Day for Disaster Risk Reduction'. Online at: www.gdnonline.org/resources/tsunami%20-genderandrecovery.pdf (accessed 2012).

DMN (2007) 'Tendances des Caractéristiques Climatiques au Maroc, Direction de la Météorologie Nationale, Exposé de Ministère de l'Aménagement du Territoire', de l'Urbanisme, de l'Habitat et de l'Environnement, Royaume du Maroc.

Ez-Zaher, A. (2008) 'Etude de la Vulnérabilité Socio-économique aux Changements Climatiques et l'Elévation du Niveau de la Mer (cas du littoral de Saidia-Ras Elma et de la commune rurale de Boudinar la)', *Mémoire de 3rd cycle*, ENFI, Salé.

FAO (2003) 'Politique de Développement et Limites Relatives au Genre', Département du développement durable de la FAO (archives).

Faouzi, N. and Khattabi A. (2009a) 'Approche Genre dans le Projet ACCMA, Analyse de la situation actuelle', Rapport élaboré dans le cadre du projet ACCMA financé par le CRDI/DFID, 2007–2010, ENFI, Salé.

Faouzi, N. and Khattabi, A. (2009b) 'Vulnérabilité des Femmes aux Impacts du Changement Climatique dans le Littoral de Nador Saidia', Projet ACCMA, ENFI, Salé.

Faouzi, N. et Khattabi, A. (2009c) 'Rapport de l'Atelier avec les Agriculteurs de la Communauté de Boudinar sur les Mesures d'Adaptation au Changement Climatique', Rapport élaboré dans le cadre du projet ACCMA financé par le CRDI/DFID, 2007-2010. École Nationale Forestière d'Ingénieurs, Salé.

Fisher, J. (2006) 'For Her It's the Big Issue: Putting Women at the Centre of Water Supply, Sanitation and Hygiene'. Online at: http://wash-cc.org/pdf/publication/FOR_HER_ITs_THE_BIG_ISSUE_Evidence_Report-en.pdf

Gurung, J.D., Lubbock, A., Hartl, M. and Firmian, I. (2006) *Gender and Desertification: Expanding Roles for Women to Restore Drylands*, International Fund for Agricultural Development.

Haut Commissariat au Plan (HCP) (2004) 'Recensement Général de la Population et de l'Habitat' (RGPH). Royaume du Maroc. Online at: http://www.hcp.ma/

Intergovernmental Panel on Climate Change (IPCC) (2007) 'Climate Change: Impacts, Adaptation and Vulnerability', *IPCC Working Group II Report, Chapter 19*. Online at: www.ipcc.ch/pdf/assessment-report/ar4/wg2/ar4-wg2-chapter19.pdf, accessed 2012.

Khosla, P. and Pearl, R. (2003) 'Untapped Connections: Gender, Water and Poverty', Online at: www.wedo.org/library.aspx?ResourceID=1

Lambrou, Y and Nelson, S. (2010) 'Farmers in a Changing Climate: Does Gender Matter? Food Security in Andhra Pradesh, India', Food and Agriculture Organization of the United Nations (FAO).

Lambrou, Y. and Piana, G. (2006) 'Gender: The Missing Component of the Response to Climate Change', Online at: www.fao.org/sd/dim_pe1/docs/pe1_051001d1_en.pdf (accessed 2012).

Masika, R. (2002) *Gender, Development and Climate Change*, Oxfam Publishing, London.

Ministry of Agriculture (MoA) (1996) Données Générales sur l'Agriculture Marocaine: Recensement Général de l'Agriculture.

Mitchell, T., Tanner, T. and Lussier, K. (2007) *We Know What we Need: South Asian women speak out on climate change adaptation*, Action Aid, London.

Mitchell, T., Haynes, K. *et al.* (2008) 'The Role of Children and Youth in Communicating Disaster Risk', *Children, Youth and Environments* 18(2): 254–79.

Ray-Bennett, N.S. (2009) 'The Influence of Caste, Class and Gender in Surviving Multiple Disasters: A Case Study from Orissa, India', *Environmental Hazards* 8(1): 5.

Rifai, N. (2007) 'Contribution à l'Elaboration d'un Aménagement Intégré du Littoral de la Province de Nador (cas des communes de Beni-Chiker et Boudinar)' *Mémoire de 3rd Cycle*, ENFI, Salé.

Roncoli, C. (2006) 'Ethnographic and Participatory Approaches to Research on Farmers' Responses to Climate Predictions', *Climate Research* 33: 81–99.

Sabhi, Y. (1998) 'Etude des Tendances des Contaminations des Organismes Marins de la Côte Méditerranéenne Marocaine par les Métaux Lourds (Hg, Cd, Pb, Cr, Ni, Ti, Zn et Cu): Aspects Environnementaux et Expérimentaux', unpublished thesis, Faculty of Science, Dhar El Mehraz, Fès.

Tschakert, P. (2007) 'Views from the Vulnerable: Understanding Climatic and Other Stressors in the Sahel', *Global Environmental Change*, 17(3–4): 381–96.

WEDO (2007) *Changing the Climate: Why Women's Perspectives Matter*. New York.

9 The effects of changes in climate and water resources on gender inequality in Boudinar community in Morocco

The case study approach

Abdellatif Khattabi[1], Soumaya Ibrahim Huber[2], Naima Faouzi[3] and Manar Matah[4]

Introduction

Despite decades of agricultural research and development efforts, the challenges of poverty and gender disparities in access to resources, markets and technologies, are still persisting. 'This persistence highlights that existing approaches to integrating gender into agricultural research don't go far enough. Many approaches aim to close gender gaps without understanding or addressing their causes', Kantor (cited in Holmes 2012) explains. The persistence of poverty, hunger and gender inequality calls for a change in the way we integrate gender in agricultural research and development. This is because conventional efforts that fill gender gaps, but do not address the underlying causes of existing inequalities, are not sufficient. Accordingly, there is a call for transformative interventions that address the social norms and power relations causing disparities in access to resources. This is along with markets and technologies, innovative organizational processes and activities that ensure gender research as standard practice in agricultural research and development (WFC 2012).

This research is a reply to that call. It is research which harbors exemplary case studies that report the data in a way that transforms a complex issue into one that can be easily understood. It adopts a social relational approach that involves placing women and men in their wider social settings, not only their households, but also their community and portrays how rigid structures impinge on the gender activities, as well as the factors affecting these.

The paper explains how the case study method was used and then applies the method designed to various households in the rural commune of Boudinar, Morocco, to examine why women are more vulnerable in agriculture and examines how external factors influence this gender relationship. This report is a narrative description of case studies presented to showcase the uniqueness and complexities of the context.

The paper is composed of five main sections. The first section is the introduction and covers the basic overview of the problem to be discussed, states the research questions guiding this work and their importance and development implications and ends with a brief description of the study group. The second section describes and justifies why the case study method was chosen as the approach and presents the five main steps adopted to carry out this work. Section three portrays the actual four case studies, each as a story, which presents the concrete narrative detail of the actual living conditions of men and women in Boudinar. It is descriptive in nature and attempts to give as much context as possible. The section is composed to handle each case as a separate case study or a story and treats each case as a chronological recounting. Section five presents the discussion and starts with the explanations of how the findings compare with other findings in the literature and whether the results confirm previous results or create a new understanding of the problem and ends with the way forward, presenting new, emerging questions.

The fundamental research questions in this study that will address the root causes of the existing gender inequalities, as described in the background, follow the general principles of the case study research questions which begin with 'how' or 'why'. They are as follows:

The first research question is: why are women more vulnerable? This led to the interview questions: 1) What are the gender division of roles and responsibilities in water management from the different sources? And 2) what are men's and women's access to and control over resources?

The second research question is: do external factors such as climate change impact on women's livelihoods and gender relations in water management in agriculture? The following interview questions were: 1) what are the effects of changes in climate and water on gender relations? And 2) what are the adaptation strategies of the family? Accordingly, the research questions conforming to the cases study principles of the 'why' and 'how' were used, as they enable the understanding and indeed challenge the root causes of gender inequality. In this sense this study is not a conventional research effort that addresses the consequences of gender equality. Its developmental implications lie in its attempt to improve gender equity, by revealing that it is the challenges ingrained in societal beliefs that limit what women can and cannot do or the assets they can and cannot hold. It thus contributes to the record to close the gender gap with consequent opportunities to sustain agricultural productivity. Thus this research should be understood in the context of the wider call for a transformational approach to gender.

The case studies

The four cases studies presented here have been selected from a pool of 28 other case studies undertaken in Boudinar community: one of the eldest communities in the region located in the northern part of Morocco, west of Nador Province. The Boudinar community is considered as the mother community for the other four communities which together compose the Tamsaman tribe; namely Tamsaman, Trogout, Awlad AImghar and Beni Marghaneen. The four cases, two men and two women, have been selected from the villages of: *Akhchab Amghar,* Boudinar haut and Abalkhach, three of the 13 villages composing the whole commune of

Boudinar. The four cases have been selected as they represent the various segments of society such as; sex, age, social status and class.

Methods

The method adopted here is the case study which is a detailed qualitative investigation of a single individual or group. The defining feature of a case study is its holistic approach, meaning it aims to capture all of the details of a particular individual or group, relevant to the purpose of the study, and living within a real life context (NCTI 2012). The case study method is defined as an empirical inquiry that investigates a contemporary phenomenon within its real-life context; when the boundaries between phenomenon and context are not clearly evident; and in which multiple sources of evidence are used (Yin 1984: 23).

While reports on case studies from many disciplines are widely available in the literature there are critics of the case study, who believe that the study of a small number of cases can offer no grounds for establishing reliability or generality of findings. In this study however, the case study method was found very useful and successful as an exploratory tool of real-life situations, issues and for generating critical findings.

The case study methodology was chosen here for several reasons: a) its descriptive, explanatory, or exploratory purposes (Yin 1993); b) it enabled the adoption of two distinct case study designs: the single-case study design and multiple-case study design. Single-case studies are just an examination of one individual or group and multiple-case studies use replications, which is despite the deliberate process of choosing various different cases, they are still likely to show similar results. This in turn helped to examine how 'generalizeable' the findings may be. The participant pool of case studies was done on a small number of cases, whereby each case focused on one participant each.

c) Further, the case study approach was adopted because it excels at facilitating an understanding of a complex issue and adds strength to what is already known through previous research. d) Because case studies emphasize detailed contextual analysis of a limited number of conditions and their potential influence, they were found useful to help in understanding the impact of climate change on the contemporary, real–life conditions and on gender relations.

The well-known case study researchers such as Robert E. Stake (1995), Helen Simons (1980), and Robert K. Yin (1984 and 1993), who have written about case study research, suggested techniques for organizing and conducting the research successfully. Drawing upon their work, the case study research followed the following steps:

- Determine and define the research questions (as mentioned earlier).
- Select cases and determine techniques for data gathering and analysis.
- Prepare for the data collection.
- Collect the data in the field.
- Evaluate and analyze the data.

Determine and define the research questions

The research began with a review of the literature to determine what prior studies have been conducted on the issue of gender inequality and climate change, particularly in Morocco. Using the general focus of the study, the research questions were formed about the situation and the specific purpose for the study was determined. The research object in the case studies was defined as the group of households (or individuals) from among the Boudinar community. The research then investigated the object of the case study in depth, selecting a variety of data gathering methods to produce evidence that leads to understanding of the case and answers the research questions.

Select cases and determine techniques for data gathering and analysis

During the design phase of the case study research, it was determined to use the approaches of single or multiple real-life cases to examine in depth and other data gathering instruments. It was decided to use multiple cases, whereby each case is treated as a single case and its information contributes to the whole study. The cases were selected because they were either unique in some way, or considered typical, or represented a variety of geographical aspects of Boudinar, and according to parameters such as female–headed, nuclear or extended families, as well as migrants. A useful step in the selection process was to repeatedly refer back to the purpose of the study in order to focus attention on where to look for cases and evidence that will satisfy the purpose of the study and answer the research questions addressed.

A key strength of this case study method are the multiple sources and techniques in the data gathering process. The research determined in advance what evidence to gather and what analysis techniques to use with the data in order to answer the research questions. Data gathered was normally largely qualitative, but included some quantitative data, for example changes in the cost of water, income or harvests, to obtain as complete a picture of the participant as possible. The case study research employed a variety of tools for data collection, namely interviews, field studies and participant-observations.

The data collection process followed a funnel-like design resulting in less data gathering in later phases of the study as the information began to be repeated, which in turn confirmed the findings during the analysis phase.

The research used the designated data gathering tools systematically in collecting the evidence. The design of the case study selected ensured that the procedures used were well documented and can be repeated with similar results relevant to the context at the time over and over again.

Prepare for the data collection

For the exemplary case studies, a clear framework with procedures was prepared before the field work. Prior to this, a pilot study was conducted in the field with four cases, in order to provide a solid design and remove potential barriers and problems. Guidelines for field procedures and data collection using multiple techniques were prepared and used. The preparations also included exercises on asking questions and interpreting answers.

Field data collection

Investigators first arranged for official permission to conduct a field work study involving interviewing the local population. The researchers then started collecting and storing the multiple sources of evidence systematically, in formats following the guidelines prepared beforehand. The investigators took written notes and recordings during interviews which were transcribed into field notes after interviews were completed. Although the interviews were open-ended, they were structured around the research questions defined at the start of the case study.

Evaluate and analyze the data

The data was analyzed holistically. Holistic analysis does not attempt to break the evidence into parts, but rather to draw conclusions based on the findings as a whole. Thus inferences were made from entire sections of the case studies rather than searching through the transcripts to look for isolatable characteristics (CSU 2012).

Thus for each individual case study, we studied its written documentation and response data as a separate case to identify unique patterns within the data for that single case. Detailed case study write-ups for each individual person met, categorizing interview questions and answers and examining the data for within-group similarities and differences were then prepared.

We examined the raw data in order to find linkages between the research objective and the outcomes with reference to the original research questions. The quantitative data collected was used to corroborate and support the qualitative data which was most useful for understanding the rationale underlying relationships.

A brief cross-case analysis followed, where we examined the cases, categorizing the similarities and differences. Patterns began to emerge, which were then tied as evidence to the findings and their relationships in answering the research questions.

Finally the 'multiple investigators' technique was adopted: the ones tested during the pilot phase were compared with the later ones in order to gain a variety of perspectives and discover the patterns. The multiple observations that converged increased confidence in the findings.

In all cases, we treated the evidence fairly in order to produce analytic conclusions answering the original 'how' and 'why' research questions.

The case studies[5]

Hassan

Livelihoods

Hassan (78) is a married farmer who lives in the village Akhchab Amghar and heads a family of eight members. The family's main source of income is derived from agriculture as well as the seasonal work of the eldest son. The father owns in his name four plots of irrigated land and eight plots of 'bour' or rainfed land. The harvest collected from their fields is primarily intended for home consumption and the rest is sold in the market ('the souk'). For several years, his net revenue from agriculture is not significant, the harvest barely manages to cover actual expenses. Neither Hassan nor any of his children have gone to school due to their meagre resources. Hence they lack skills, and cannot perform any other activity apart from agriculture. In addition to crops, he raises some livestock, namely six sheep, two cows, two kids, three goats and two lambs. On the whole his net total revenues from agriculture do not exceed 5000 to 7000 dh[6] per year. His wife raises some poultry, which is mainly for the family's own consumption, but can sometimes sell some of these.

Gender division of roles and responsibilities

All reproductive roles of cooking, dish washing, housework, laundry, bathing, are done with spring water and performed by females of the household, who are also responsible for the transport and saving of water. Productive roles are performed by both males and females, who practice traditional agriculture. Females, for example, are always responsible for the preparation of the field, for sowing the seeds of wheat and barley, cleaning crop produce in the river, or at home, and then transporting it back home on donkeys.

Male members are always responsible for some activities, such as: ploughing using an uncle's horses, sowing wheat and barley, repairing water courses and irrigation channels, which are often damaged by the flood of the Amekrane River. The seguias (irrigation channels) are sometimes repaired by the authorities or farmers. Sometimes both males and females are responsible for picking the fruits as well as performing some gardening activities.

Animal husbandry is entirely a female task. Females of his household are responsible for feeding animals, getting drinking water from the spring and cleaning the livestock locations. Thus in addition to household chores, his daughters are involved in agricultural work as well as animal husbandry. The daughters bring back the forage from the field, clean the barn and take care of the cattle during periods of rain. Selling livestock in the market, however, is confined to males, while selling poultry to the neighbors is the women's role.

Access to and control over resources

In general men have full access to and control over all financial and real estate resources. They are involved in transactions for sales and purchase of agricultural products or livestock and manage the revenues generated. Women do not participate in this process.

Although Hassan's daughters are in charge of much of the work on the land as well as the livestock as presented, they receive neither information nor monetary compensation for their work. All decisions are entirely made by the father, such as the choice of seeds, grains, pesticides; setting the price for the harvested products and all other issues related to the management or operation of the land.

His wife and his young daughters possess no source of income to cover their personal needs, and hence sell their poultry products to neighbors. As all the other women in Boudinar, his female family members never go to the souk. Hassan and his daughters practice traditional agriculture and hence do not own any agricultural technology.

Effects of changes in climate and water on gender relations and their adaptation strategies

The village Akhchab Amghar, where Hassan lives, is one of the thirteen villages of Boudinar and hence its problems, traditions and customs and living standards are similar to those of the other places within the commune. Its only peculiarity is that it is relatively accessible as it is closer to the center of Boudinar, although when river flooding occurs, it is also difficult to reach. The main activity for its inhabitants is subsistence agriculture. There are no other job opportunities with the exception of a few opportunities related to building construction and the collection and cleaning of sand in the river to sell it for entrepreneurs.

Collecting water from wells and springs at the village Akhachab Amghar is free of charge. Currently, as the last two years brought much rain, water is available in springs and women are well served with water to perform their tasks.

Now as before, men retain access to and control over all financial and real estate resources. They are involved in all types of transactions for sales and purchase of agricultural products or livestock and also the management of the money generated through commercial transactions. Women do not participate in this process.

In the past, the harvest was always sufficient to meet the needs of the family. At present, but in rare cases, when the harvest is not sufficient, the father resorts to the sale of livestock, and the mother and her daughters sell their poultry to their neighbors. Job opportunities have become scarcer except for the few opportunities in the building sector, in transportation and harvesting of sand from the river. Today, due to personal reasons such as aging and the health issues of Hassan, a small part of the land is planted and the rest is left abandoned.

For Hassan the main specific problems related to water or climate change are: a) storms and droughts, because they cause crop loss and b) plant diseases as they result in loss of harvest and increased use of pesticides.

Ahmed

Livelihoods

Ahmed (50) is an illiterate farmer, who is married and lives with his 10-member family in the village of Abalkhach. Ahmed owns two plots of land: one in irrigated agriculture and another in the 'bour' or rainfed agriculture. The harvest of both lands is primarily for the family's consumption and also for marketing purposes. On this land they grow wheat, potatoes, beans, watermelons and melons. They have also 500 olive trees for the production of oil. According to Ahmed, the planting of fruit trees is better suited to the region and involves less work. The harvested products from agriculture yield a net revenue of 15,000 dh per year, which is sufficient to meet the needs of the family.

Gender division of roles and responsibilities

Ahmed's household uses water from the neighboring river during summer for all household chores, such as cooking, dish washing, house cleaning, bathing, toilet and laundry. It is the female members in his household who are responsible for all these reproductive activities. Transporting and getting rid of waste water is also a female activity, but purchasing potable water is done by the male household head. Both male and female family members can sometimes be found to clean the water cistern. Ahmed's daughters collect the drinking water once per day from the spring, which is a half an hour walk from home. In summer they collect water from the spring several times a day. Often during summer water is purchased with 150 dh per week to fill in the tank, which amounts to at least 600 dh per month.

Ahmed's daughters participate in all reproductive as well as productive activities. Men are always in charge of irrigating the land with water diverted from the river, repairing the irrigation channels and protecting agricultural lands from the flood by setting up terraces on the sloped land. Harvesting is an activity that is always done by women and seldom by men. Sometimes female family members can participate in work protecting land from the floods and men can also contribute to washing the harvested products, though the latter is usually performed by women. Washing the harvested products in the home however, is always a women's activity. Animal husbandry is always an entirely female responsibility; they feed the animals and clean their locations.

Access to and control over resources

All agricultural activities are performed by his daughters and his son; together they clean the land, do the sowing, harvest and collect agricultural products. The sale of the harvest, however, is undertaken solely by his son, who is responsible for the management and control of the resources as well. He is in charge of all commercial transactions, seed selection, and all decision making regarding farm management. The land was owned by Ahmed's deceased father, and it is his son who succeeded him, who continues to operate it now. His daughters can practice agriculture as long as a male member of the family is present in the village and participates, even if only symbolically. His daughters hold as much knowledge and experience in agriculture as his sons, and some tasks are done only by his daughters, such as harvesting and cleaning of the harvest, and maintenance of livestock. Farming is not an important activity in the village, and the little that is done is primarily for the household subsistence.

Effects of changes in climate and water on the gender relations and adaptation strategies

In recent years, due to erosion and flooding, the family stopped field seasonal crop agriculture and moved more towards the planting of fruit trees including olive trees. Even though this change reduced substantially the amount of agricultural products intended for home storage, the olive trees remain relatively good and generate a reliable income. No precaution, however, is taken to ensure the proper conservation of the stored agricultural products that degrade very quickly because of air humidity.

Nevertheless, the expenses constitute an important concern in the family's finances: first of all because of the losses and damages caused by the floods and erratic rainfall, and then because of the increase in price of fuel and other products such as fertilizers and pesticides, and finally as the labor force has become scarce and expensive. While the expenses remain high, the prices of agricultural products in the market have decreased. Farmers sell their products at low prices, because they do not have the means to store and retain them. The purchasing power of local people has also dropped due to the economic crisis and the sudden losses due to the impact of climate change. Similarly the financial aid coming from abroad has also decreased dramatically as migrants have also encountered economic difficulties in places where they live.

Changes in the water sources

Indeed, all of the villages are devoid of basic infrastructure, roads are defective or non-existent and only three out of 13 villages benefit from piped potable water. Households resort to the use of the water collected from wells and springs. The collection of water, which always remains the responsibility of women and children, occurs more easily, with less waiting and less conflict during rainy years. The purchase of water occurs only during the summer period, when sources dry up, and access to water gets difficult and this happens generally during drought years. As the waiting time to collect harvest water gets longer, conflicts arise between the inhabitants (the last to arrive are less well served). During winter time water becomes more available and women can easily access the water sources.

The water situation is exacerbated through the isolation that the area faces during rainy periods, as well as the lack of transport vehicles to get to the village centre or the neighboring cities. In recent years, the 13 villages that make up the rural commune of Boudinar get separated by the Amekran River and its heavy floods, as they are each located on different heights. The rain is always beneficial as it feeds the underground water table and hence some wells and springs that have dried out for a long time get revived. The availability of water has allowed for a relative development of agriculture in the region and also reduced the farmers' expenses for purchasing potable water to fill up their tanks, especially during the summer season when water gets expensive. For girls especially, water collection from sources endowed with much water make it easy and fast for them and thus the time spent fetching water is reduced. However, the torrential rains are usually accompanied by strong winds which break the irrigation channels and trees in the farms. The region often also faces seismic events that still endanger some water sources and facilities especially for drinking water.

In addition to these natural and economic problems, there are scanty state interventions to fully address these problems. Even though the authorities may be aware of these issues, they still lack enough material and financial resources necessary to address fully the encountered problems. There are almost no other alternatives available for income generation. Men can, on rare occasions, get involved in building construction activities, gathering and cleaning sand from the river, and some small businesses trading various goods in the market.

To avoid some of the problems caused by climate change, including plant and tree diseases and loss of production, many people have abandoned agriculture or changed the type of plants they farm. Ahmed for example, because of flooding, has shifted to the planting of fruit trees, such as olive trees and fewer cereal cultures. This has meant an increase in women's tasks as they are the ones involved in harvesting olives. There is however a withdrawal of men, who prefer to work in building construction or migrate to large cities.

For Ahmed, the main problems related to changes in water and climate are, in order of importance: a) frost or cold days, as they destroy the seeds and plants, b) wind, because it knocks the plants, c) droughts, because they create water scarcity and d) high intensity rain storms, which ruin the harvest and erode the soil.

Changes in gender roles

The distribution of gender division of roles and responsibilities in Boudinar has varied and changed over the years, which has also meant a change in the status of women, socially, economically and culturally. This change, however, was directly related to either men's absence or presence in the community as well as their participation in economic activity. Women have often been used as an overriding element to perform the tasks which are usually performed by men, such as irrigation, plowing and transportation of agricultural products. This development of course allowed women more access to and control over the resources during times of men's absence.

Before the 1960s, for example, as some local persons have mentioned, women were traditionally granted more freedom and could move easily in the village, have access to the market and carry out commercial transactions, such as the purchase or the sale of products. The drought of the 1970s greatly impoverished villagers, some of them had to give up agriculture and immigrate abroad or to some big cities looking for a job. Because of this important migratory flow, the villages were abandoned by a lot of their male population. Women and girls, who were left behind, were bound to remain in the villages.

Thus, because of their absence, men have imposed movement restrictions on women, not allowing them to go out, hence it has now become a tradition and married women cannot go to the market or work publically in agricultural activities. Now young girls are involved in agriculture, in addition to the collection of fuel wood and drinking water. Although migration flows today are almost non-existent, young men have become accustomed not to participate in agricultural tasks.

The role of the state

The state did not integrate in its forecast any specific activities for female agricultural workers and those who have no means other than agriculture to meet the needs of their families, including women heads of households. Indeed, fruit tree planting actions carried out by the state under the Green Morocco Plan to overcome the problem of erosion, have helped some farmers and women to overcome their difficult financial situations, especially if their only resource is agriculture, as harvesting fruit trees is primarily a female occupation. Another gap is also reflected on the education level of girls and boys of different villages, where the illiteracy rate is very high and where school absenteeism is a real problem. Both young girls and boys, generally, have to leave school to help their families in agriculture and with house work. The lack of secondary schools in the villages and the absence of paved roads and of transportation means are also a part of the reason why young boys leave school at the primary level of education. There are few young girls who continue their studies up to the end of primary school. Usually lack of security and traditional customs often necessitate that girls leave their schools at an even younger age.

Aziza

Livelihoods

Aziza (43) is an illiterate housewife, who lives with her husband and their five children in the village Boudinar-haut. Her family's source of income is agriculture in dry land farming, which the family uses for its own consumption as well as for marketing. The family cultivates carrots, turnips, potatoes, wheat and barley. The agricultural revenues, however, do not cover the needs of the family. To close the gap, her son works as a seasonal worker in buildings construction and agriculture. Recently Aziza bought some chicks for 12 dh each, which she raises in order to sell them a year later for 100 dh a piece.

Gender division of roles and responsibilities

As a normal housewife, Aziza is responsible for all the reproductive activities and is assisted by her daughter. She never goes to the fields, but participates in the agricultural work in the land parcel adjacent to her house.

Aziza uses tap water for all the house activities, such as dish washing, laundry, bathing, and cleaning of the house. The water is usually reused for some household maintenance purposes as well as for the toilet. Aziza is the one responsible for storing and throwing out of waste water. The males in her household are responsible for purchasing and transporting of water.

The water supply has not been impacted by climate change as it is provided through a piped network. All households have benefited from the supply of piped potable water, but water is not always available and there are often cuts of water especially during the summer period. As all Aziza's household activities are using water from the tap, the family resorts to the collection of water from springs only during the times of water cuts. The water is often available in springs and in the river, and waiting time during water collection is therefore shorter. For the needs of her family, Aziza and her daughter collect water from the spring three to four times per day, and each trip takes 30 minutes. In the summer, when the water is not much available in the river or in the spring, the family uses tap water to clean the harvested vegetables, which consequently raises the water bill to 250 to 350 dh per month.

Her daughter participates regularly in all agricultural activities, such as weeding the fields, sowing, and harvesting agricultural produce. Male farmers are always responsible for irrigation, repairing of water channels and protecting the land from floods. Sometimes as well her daughter and other female members of the family help with irrigating the fields. Washing and cleaning the agricultural products (vegetables) on the land is usually performed by both males and females. Washing the harvest, however, at home is an entirely female activity.

Animal husbandry is usually a female responsibility. Women wake up at 7 am to clean the barn where the livestock stay. Feeding the animals, however, can sometimes be taken over by children in the household. Men are the ones in charge of grazing the animals in the camps.

Access to and control over resources

All agriculture related decisions are taken by the father; he takes care of the sale of the crop and the management of the resources and decides on the types of grains and agricultural inputs as well as the sale price of the harvest. His wife Aziza can be informed but can never access revenues. The fruits of the harvest are transported in a pickup truck by the father to the market and the revenues are used to meet the needs of the family and agricultural inputs. Aziza has never received any credit. In case of need, she requests the support of her friends or family. For the family's travel, the father rents a car.

Livestock related decisions, especially small ones (sheep, poultry and rabbits), are taken by her, the mother, as she uses their products for the household's consumption. In case of need, her husband sells the sheep in the markets and she sells the poultry and rabbits in the neighborhood, and hence can benefit from that money.

Effects of changes in climate and water on the gender relations and adaptation strategies

In previous times, Aziza used to collect the water on a donkey, and wash the laundry and the harvest at the river. Currently, and since there is a supply of drinking water by the connection to the municipal water network, everything is done in the home, even washing the agricultural harvest and the laundry. Women resort to the collection of water from outside (springs and wells), only during water cuts, which are recurrent. The time, however, allocated for this task is relatively short. Fetching water for the animal husbandry has also become easier so that women work less at night now and sleep a little longer in the morning.

On the other hand, working in the fields to remove weeds has increased. Because of the floods, however, farmers suffer big losses, as the irrigation channels are destroyed and the harvests are spoiled. To offset these losses, her husband and sons go to work in the building construction sector as daily workers and abandon agriculture and it becomes women's duty to repair the damage.

To support her family during periods of rain, it is her duty as the mother to store a portion of the harvest, which can sometimes also be marketed if needed. The storage is done randomly.

The family's eating habits have experienced some changes; nevertheless, the potato is still the main food because it is easily storable. Because of the cold and humidity, family members often suffer from colds and infectious diseases during the winter season. Her eldest son often gets sick because of cold, which prevents him from working.

Aziza, like the other inhabitants, suffers from the decrease in the quantities and prices of the harvest, due to the loss of crops and the decrease in the purchasing power of communities. While men abandon agriculture for work in the building construction sector there are fewer agricultural activities available for women.

For Aziza, the main problems related to water and climate change are constituted in the difficulty of accessing health centers and purchasing drugs in case of flooding. Due to the cold and humid weather, her children are often sick with infectious diseases. The problems with regard to agriculture depend not only on the availability of water, but also on the increase in prices for seed, fertilizer, fuel and transport systems.

Aziza's family has no access to any information on climate change and they don't know any active cooperative or association in the municipality that could possibly help them out. The adaptation strategies of Aziza's family to manage budget deficits and meet food insecurity with lower harvests compel her husband to work as a daily laborer in building construction. She resorted to raising chickens or rabbits to meet the needs of her family as well as for marketing purposes.

Louisa

Livelihood

Louisa (30) is an illiterate and single seasonal farm worker, who lives together with her small family composed of a sister and mother in the village of Boudinar-haut. Their main source of income is from their seasonal work in agriculture, from which she earns 30 dh to 40 dh per day during regular working days and 50 to 70 dh per day during agricultural products harvesting time. On average she works 15 days per month. Louisa is the main income provider for her small family and her mother has been sick for many years and depends on her support.

Gender division of roles and activities performed with different water sources

Louisa is responsible for all reproductive activities inside her home. For all household chores such as cooking, washing dishes, bathing, cleaning, she uses tap water. Washing the laundry is done in the river. They do not, however, have any tank to store rain water. Louisa is also responsible for getting rid of the waste water. After her work in agriculture, and upon her return from the fields, she collects firewood in the fields once every three days. This firewood she uses for cooking, bread baking and water heating. Louisa is as well responsible for the housework maintenance, such as repair and construction of roofs and barns.

As for the productive activities, Louisa participates as a daily laborer in all agricultural works except for irrigation. There are some other female laborers like her, who are engaged as manual laborers on farms. Usually the farms resort to their own family members for help in the fields. The tasks she performs as a female laborer are soil preparation, repair of water channels and washing the harvest in the river. She carries the harvest either on her back or on her donkey. It is primarily the male workers' responsibility in the fields to do the irrigation and repairing of irrigation channels.

Animal husbandry is also Louisa's sole domain. She is always the one to feed her animals, clean their places, and repair and construct the barns.

In the area of community management, Louisa rarely invests much time. Yet in a few cases after the floods she helped other community members in the conservation and rehabilitation of their houses.

Access to and control over resources

On the land, where Louisa works, the land owners are the only ones who are in control of everything, because they have the land title in their names. They are the ones who decide on the type of seeds and agricultural inputs and purchase them from the market and arrange for prices of products. Louisa is there only as a worker, who plants the seeds and harvests the product. Through her work on the land, however, she and her sister as well as the other workers have access to the traditional knowledge about techniques. Louisa for example gets access to information about adaptation techniques. Louisa receives credits from her grocery owner in order to buy her food. She does not have access to any education and has no means of transportation.

Effects of changes in climate and water on the gender relations and adaptation strategies

Louisa has witnessed several changes with regard to the water sources. In earlier times, Louisa, like other people, had used the spring water. She and all the other women woke up very early to fetch the water which was a 30 minute walk from her home. When it rained, the roads became impassable, and it was difficult for her to go to fetch water, as she was landlocked.

Now with the recent changes and the introduction of water pipelines, a change occurred in the cost of water. While in the past water was free, since the village has been supplied with piped drinking water, she pays 30 to 40 dh per month. Louisa is the sole person responsible for paying the cost. The laundry is done at the river to reduce the monthly water charge.

The introduction of water pipelines has also changed her use of time. Previously she had to go fetch water four–five times per day (four cans of five liters). The first time was very early in the morning so as to be able to focus her entire attention later on her work. Now she can spend more time with her ailing mother and in particular earn some money. Previously, she used also to interrupt her field work during the day and go back home to do some work and care for her mother.

Closely related to this is also the impact climate change had on her workload. Due to the massive precipitations, she has to repair water channels (seguias) in the field, as well as her home, which is built of adobe, and is hence very vulnerable to rainstorms. Thus, she must evacuate water and repair the damaged roofs and walls. After heavy rain storms, Louisa and her sister spend the night in the drainage of the rooms and in the morning they start rebuilding what was destroyed by the rains. The activities related to home maintenance are expected to be performed by women. Sometimes she and her sister work in roof construction and repair other homes and get paid for the service.

Over the last years, work opportunities available in the community have decreased drastically. Due to the floods, water channels were destroyed and the number of plots available for farming dropped. Indeed, several people following the floods have abandoned agriculture. Thus in agriculture, labor opportunities are very little now as there is less harvesting and gathering. Many who did not give up agriculture completely started planting fruit trees, but only in the barren dry land. In addition, farmers are more and more investing in livestock. As Louisa's income from agricultural work dropped because of the recurrent flooding, she started as well to raise sheep and poultry to improve her income. While her agricultural activity decreased, much of her time is devoted to the evacuation of water from the houses and rooms.

In terms of food security, Louisa's household has also witnessed changes. They now consume more legumes and vegetables and less meat and couscous, and pasta substitutes bread. During rainfall, the family eats mostly pasta and couscous, instead of bread, due to lack of fuel wood. Due to their meager livelihood means, they do not possess facilities for storing agricultural products. The family depends only on their purchases from the grocery shop as she never goes to the market.

The health status of her family members has also been impacted by the climate change. Because of the recurrent cold weather and humidity, her mother suffers from rheumatism. The mother is often bedridden and can no longer participate in the work of any of the reproductive activities. This of course increases the workload of her girls and impedes their working opportunities in the fields, as they need to attend to their mother. Needless to say they have fewer resources as a result.

Due to these cold winters, which are getting ever colder as well as the lack of means of protection, people get sick more often and work in the fields becomes very difficult for them. Sometimes, there are periods of high rainfall or lasting droughts, when the agricultural activity stops completely and some people even abandon farming. In the summers, on the other hand, it gets very hot and work during the harvest becomes very difficult.

In general the work has decreased in agriculture and there are also several other changes noted in the agricultural resources. For example now there are more fruit trees, as they require less investment and less water. Men are not as heavily involved as women in that. Even though the female labor has increased, they continue to be paid less than men. Sometimes women are paid in kind, for example through a few liters of olive oil. In this they have no control, as pointed out before.

The climate changes did not result in any conflicts over water or land title among family members. For Louisa, the main problems occurring due to the changes in water and climate are due to the floods, as they result in destruction of their houses, isolation of the area and loss of crops, or droughts that impact on agricultural production and hence the opportunities for work.

Coping and adaptation strategies (present and past)

Louisa has neither access to any type of institutional support, nor to any type of information about the latest changes in water or climate change. To cope with the shortages of income, Louisa's family resorts to the water available at the water sources so as to save the cost of the tap water. The saved money enables her to buy food from the grocer and do some poultry production. During floods Louisa disposes of rainwater.

In order to maintain themselves, the two sisters earn their living through casual and low paid labor. They work as agricultural workers, they also serve in private households as maids and during summer they get involved in some construction work such as repairing roofs for urgent work, caring for animals and moving furniture.

At present their work opportunities are limited to agriculture and household serving. This is because previously the houses were built of adobe and hence after the rains there was a lot of repair work that needed to be done. Now people are building concrete houses, which requires less work.

Louisa observed that none of the techniques for water harvesting, storage, carrying and saving had changed. For agriculture, some people use tractors for tillage and combine harvesters to harvest. Because such technology requires a lot of investment and as the land plots are small due to the recurrent floods, it is only used by a few farmers.

Over the past three years, and with the increased availability of water in wells and springs due to regular precipitations, farmers resort to pumping water from the river and wells. They use pipes to bring water from springs and from the river to the fields. They also use the water channel system which is often damaged by floods and the overflowing of the river. Especially during the winter, water is abundant in the river for irrigation. In summer, there is less water and this creates conflicts between different users.

Discussion

The findings in this study converge with many other studies which started immediately after the publication of Boserup's book in 1970, as they frame women and men in agriculture in policy, development practice and research. They draw attention to the roles, interests and the asset bases of women and men in agriculture, and link these to gender relations. Key elements of these framings include the following ideas:

- women perform most of the agricultural work in addition to their responsibility for the reproductive work and have limited control over their own labor and time.
- women lack access to land and latest innovations.
- women's lack of control over resources such as land and education as well as mobility restrictions, prevents them from economic undertakings and limits their new opportunities, including accessing new markets in the agricultural sector.
- women's work burdens increase as a result of the out-migration of men seeking other income earning opportunities, and as access to water and fuel has deteriorated with environmental change.
- women have limited control over the outputs from their labor and therefore lack incentives to increase their production. They are relegated to the few meager income earning opportunities such as poultry.

In summary, in relation to where these findings fit into the four decades of work on women in development, the findings in the literature collectively paint a picture of rural women working in agriculture who are vulnerable, overburdened, least rewarded, and poor. Nevertheless, equally, though less evident, they play the main role in the provision of food security and the well-being of their households, primarily during the absence of the males (IFPRI 2002).

In this literature there always remained the need for more context-specific evidence that documents what women and men do in agriculture and how social institutions, including the household, impinge on these activities, in order to better define areas of intervention that do more than deliver technical inputs without addressing the wider structural factors influencing whether and how women engage in agriculture.

The findings in this study however expanded our understanding of the problem as well as created a new awareness by highlighting and confirming two main conceptual points:

First: that gender relations are dynamic, as women and men seek to maintain their livelihoods in response to their day-to-day realities in addition to changes occurring at the meso and macro levels. Unlike the orthodox understanding of gender analysis as promoted by the Harvard framework, which provides a static view of women in agriculture, and that aims at presenting women as the disadvantaged and highlights the nature of their disadvantage by focusing on time inputs, assets (especially land but also credit conditional on land access) and women's caring roles (March, Smith and Mukhopadhyay 1999), the case studies presented here provide an alternative approach. This approach began by examining the various compositions of households (meaning male-headed or female-headed) in various locations of Boudinar, and detailed how they account or how they operate in terms of livelihood and meet their responsibilities.

It then proceeded to examine their access to and control of available household resources and the implications for individual decision-making and household livelihoods that extend over time and over the reproductive and productive range of activities. Such an approach led to a different set of questions, different data, and particularly qualitative data on gender relations unlike the existing, simply sex-segregated role data (Leach, Scoones and Stirling 2010). Further on, such a shift from this kind of analysis proved useful because it did not detach women and men from their social environment, but asked instead about their own clarifications and perspectives about their changing contributions amidst the changing circumstances of their social and natural living environment. Hence it did not take gender role data as the end point of gender analysis. It is hoped that its design might also result in a more effective approach to address disadvantage and thus support a strategy for closing the gender gap.

Second: that women and men, who must be understood as diverse social groupings that encompass multiple identities as individuals, spouses, workers, parents, siblings etc. (Okali 2011a), are governed and influenced in their relations by the legitimacy of social norms. These norms are the ingrained societal beliefs that limit what women and men can and cannot do, or the assets they can and cannot hold. The limit also extends to the personal level and impregnates the mindset. This case study approach enabled a detailed analysis, which facilitated a more complex and more nuanced understanding of the relations between women and men and even of the understanding around assets control. Here the case studies were able to highlight the dynamic nature of these relations and the less obvious aspects than under the conventional framings (Okali 2011a).

Two patterns appeared in this dynamic nature. One during the first drought era in the 70s, when men migrated outside or inside Morocco and women no longer enjoyed the social norm that allowed their mobility. Hence women were not able to go to the markets and sell their products anymore. The second pattern appeared during the later phase of the recurrent floods and droughts, which led to the decrease in the quantities and prices of the agricultural harvest, and the decrease in the purchasing power of communities, hence agricultural activities decreased for both men and women. While men easily abandoned agriculture for work in the building construction sector, women were faced by several social restrictions. Current social norms restrict women's geographical mobility, and enforce as well a horizontal division of labor, by restricting their movements between working sectors or occupations.

The findings have thus highlighted that decisions taken by household members to change their economic activities depend partly on what is going on outside households and not only on what goes on within households as presented by the women in agriculture literature as the central theme.

The way forward and new emerging questions towards achieving sustainable transformative change for women

As brought forward by this study, it is essential that research efforts move away from their classical patterns of focusing on women as a bounded group or framing the rural communities as a group of isolated individuals who have their separate interests. Rather the new emerging research questions should address women and men as active social agents and place them within their wider social and dynamic contexts of gender, age, class and focus on identifying how women and men experience and value ongoing changes and use them to meet their own interests while addressing their family's concerns.

New transformative questions should investigate closer women's and men's mind sets to help them delve more into themselves and bring out what, why and how they would envision themselves as empowered. Privileging sex-segregated data or a purely production-oriented view of development over a relational and well-being oriented one will never help achieve this aspiraton. The transformative research questions should focus on the decision-making roles of men and women in light of attitudes, norms and beliefs at the level of households, community and beyond as called for in the collection of papers by Kate Young and others published 30 years ago (Young, Wolkowitz and McCullagh 1981).

These different sets of research and policy questions should of course reflect the specifics of particular locations and situations of different categories of rural women and men. The assessment should not focus on women themselves for change to be sustainable but also on attitudes of those around them within households, wider family units, and communities, simply the circumstances external to them, bearing in mind that the focus of change starts in the persons themselves who further affect their surroundings.

From her comparison of Bangladeshi women in London and Dhaka, Kabeer confirmed that women will not be in a position to either gain any sustainable advantage from participating in new opportunities, or alternatively, to protect themselves as resources become scarcer for example, if the wider social, economic, political and institutional environment is not supportive of any claims they make (Kabeer 2000). Social legitimacy is essential for realizing gendered claims, not just legal and policy support.

Most importantly, transformative research should through its questions allow women and men as active change agents to start the change process (occupation, market, policy etc.) up til the envisioned outcome even during the research practice. The development and policy implications of the research findings can be seen on two levels: a 'deeper impact' and 'wider impact'. Deeper impact because policy needs to go to the deepest inner core of the change agent and strengthen the men's and women's individual capabilities. Wide impact is a call for policy to widen its understanding of the wider social dynamic interrelations which frame the decision-making levels.

Conclusion

In this research, case study was used to explain the situation, to explore, or to describe the phenomenon of gender inequality and how it can be exacerbated through external factors such as climate change or migration. The case study method proved to be useful because of its applicability to real life, contemporary human situations and its facilitation for an understanding of complex, real-life situations. The cases used are illustrative case studies, primarily descriptive studies, which utilize four typical instances of individuals to show what the situation is like. The illustrative case studies served primarily well to make the unfamiliar familiar.

In other words, this research is an effort to reveal how climate change can affect productivity but importantly can widen the already existing gender gap due to the underlying gender norms. Productivity in this case becomes much broader than crop yields and the exerted efforts of women on sustainable food security. Productivity in these cases becomes paramount with gender equality.

In conclusion, this work is a documentation hailing the call that for any agricultural effort aiming at benefiting both men and women in Boudinar, gender equality is a prerequisite and hence needs to be understood in its complex, real and dynamic life situations, which can best be accomplished through a case study approach.

Notes

1 École Nationale Forestière d'Ingénieurs (ENFI), Salé.
2 Consultant, Cairo.
3 Consultant, Casablanca.
4 Haut Commissariat aux Eaux et Forêts (HCEF), Chefchaouen.
5 For ethical reasons the real names of the case study interviewees have been exchanged with symbolic names instead.
6 1 Moroccan dirham = 0.1172 US.

References

Boserup, E. (1970) *Women's Role in Economic Development*, St Martin's Press: New York.

CSU (Bronwyn Becker, Patrick Dawson, Karen Devine, Carla Hannum, Steve Hill, Jon Leydens, Debbie Matuskevich, Carol Traver, and Mike Palmquist) (1994–2012) *Case*

Studies. Writing@CSU. Colorado State University. Online at: http://writing.colostate.edu/guides/guide.cfm?guideid=60

Holmes, H. (2012) 'Gender Transformative Change: A Key to Lasting Agricultural Development Impact'. Online at: http://aas.cgiar.org/activity/news/gender-transformative-change-key-lasting-agricultural-development-impact#.UgiyB6Pfpjo

IFPRI (2002) 'Sustainable Food Security for All by 2020', Proceedings of International Conference, Bonn, Germany (4–6 September 2001). Online at: www.ifpri.org/pubs/books/2020conpro.htm#153

Kabeer, N. (2000) *The Power to Choose: Bangladeshi Women and Labour Market Decisions in London and Dhaka,*:Verso: London, New York.

Leach, M., Scoones, I. and Stirling, I. (2010) *Dynamic Sustainabilities: Technology, Environment, Social Justice*, Earthscan: London.

March, C., Smith, I. and Mukhopadhyay, M. (1999) *A Guide to Gender Analysis Frameworks*, Oxfam: Oxford.

National Center for Technology Innovation (NCTI) (2012) Online at: www.nationaltechcenter.org/index.php/products/at-research-matters/case-study/#_ftn1 (accessed 2012).

Okali, C. (2011a) 'Enabling rural women's economic empowerment: institutions, opportunities and participation in Accra', Expert Group Meeting, Ghana UN Women In cooperation with FAO, IFAD and WFP (20–23 September 2011) (3). Online at: www.worldfishcenter.org/sites/default/files/Okali_2011_0.pdf (accessed 2012).

Okali, C. (2011b) 'Integrating Social Difference, Gender and Social Analysis into Agricultural Development', FAC Policy Brief 039, IDS: Sussex.

Simons, H. (1980) *Towards a Science of the Singular: Essays about Case Study in Educational Research and Evaluation,* University of East Anglia, Centre for Applied Research in Education: Norwich.

Stake, R.E. (1995) *The Art of Case Study Research*, Sage: Thousand Oaks, CA.

WFC (2012) 'Building coalitions, creating change: an agenda for gender transformative research in development'. Online at:www.worldfishcenter.org/gender-research-workshop/ (2)

Yin, R.K. (1984) *Case Study Research: Design and Methods*, Sage: Newbury Park, CA.

Yin, R. (1993) *Applications of Case Study Research*, Sage: Newbury Park, CA.

Young, K., Wolkowitz, C. and McCullagh, R. (eds) (1981) *Of Marriage and the Market: Women's Subordination Internationally and its Lessons*, CES: London.

10 Social Analysis Systems (SAS) for gender research and social capital enhancement

Asmae Harrad[1], *Fatima Zahid*[1], *Maria Fernandez*[2] *and Aden Aw-Hassan*[3]

Introduction

The goal of the Project for Building Capacity in Social and Gender Analysis (SAGA) was to build capacity and generate innovative practices and methods that demonstrate the value-added of social and gender analysis methods in action-oriented and transformative research. The SAGA project has provided an opportunity for researchers to build skills with participatory methods, especially those that are used in social analysis systems (SAS). Over a two-year period (2010–11) a team of professors and graduate students from the National Agrarian Institute of Morocco located in Meknes carried out a multi-part study on the role of women in the management, use and marketing of aromatic and medicinal plants in the village of Maâmar in the middle atlas province of Khénifra.

Maâmar, like other middle atlas communities, is Amazhir-speaking and most women are illiterate. The distribution of knowledge, assets and decision-making power are appropriate to its Berber heritage. A challenge to researchers in cross-cultural situations such as this is to avoid assumptions, stereotypes and miscommunication. The choice of research methods and tools is critical if an opportunity for community members, and especially women, to express their opinions and to learn as much as possible is to be ensured.

SAS tools were used for the research with the intention of enhancing not only the quality of findings but also to reinforce social capital in Maâmar. The part of the research described here was designed to assess the impact of SAS tools on the quality of information gathered and the learning of the men and women of Maâmar during the study. This article will assess:

- the choice of tools for different aspects of the research;
- the degrees of satisfaction of the community members with the information generated using the tools.

The study will also assess the impact of different tools on women and men regarding their understanding of:

- social, economic, agro-ecological and environmental issues;
- organizational issues in the community;

- ways of reducing inequality between men and women;
- ways to enhance the voice of women;
- the context of the study.

The research was done under the umbrella of the Social and Gender Analysis (SAGA) project of the International Centre for Agricultural Research in Dry Areas (ICARDA) that was financed by International Development Research Centre (IDRC) of Canada. The goal of the project was to enhance the capacity of researchers of the Middle East and North Africa (MENA) region working on natural resource management issues, using a gender perspective. Studies leading up to the project had identified the need for a sound understanding of social, gender, economic and institutional dimensions of environment and natural resource management (ENRM) including agriculture if appropriate development interventions and policy responses were to be designed. The studies recognized that research in the region has made significant progress in tackling different issues such as poverty alleviation and the sustainability and improved production of water, land and biodiversity resources but that gender and social analysis needed strengthening. The major objectives of this project were to:

- identify and conduct three applied research projects on gender and social analysis to generate new knowledge on key challenges of gender, poverty and ENRM;
- strengthen capacities of MENA researchers to implement action-oriented research using social and gender analysis on ENRM by providing technical support to deepen current research approaches, methodologies and analysis; improve peer learning and networking among researchers in the MENA region through peer exchange, joint analysis, and capacity building activities;
- identify key issues of organizational support for integrating social and gender analysis in ENRM research through structured discussions of challenges, successes and strategies with researchers and management;
- disseminate key results, lessons and methods of the program to different user groups through communication strategies and production of media and materials.

Tools for transformative research

Social Analysis Systems

The Social Analysis Systems (SAS) was developed as part of an international collaborative project to bring power of thought to research by offering new ways to create and use knowledge for the betterment of all involved. SAS aim to unite and enrich a wide range of concepts, tools and transferable skills that allow us to stimulate creativity and human intelligence, to better respond when change is needed. The SAS manual brings together some 80 tools for this purpose in a systematic manner. It is the task of the researcher to choose a set of tools that will enhance the quality of his/her research and facilitate learning for change. The tools assessed here were chosen to enhance the research in Maâmar on the role of women's community organizations that influenced their role in the management and transformation of aromatic and medicinal plants.

Tools were chosen that would: 1) allow the identification and analysis of different organizations in the community and people's perception of them, 2) facilitate in-depth study of the relationships within and among them (history, conflicts and alliances) and 3) generate information for research that could also be used by the community to better understand and solve its problems.

Social capital

The concept of social capital provides us with a framework to evaluate the degree to which a community understands the processes it is engaged in and can make decisions for its common future. Social capital includes mutual trust, reciprocity, recognition of groups, a collective identity, a sense of a common future and capacity to work together (Putnam 1993b). Social capital is a measure of intangible assets in a community such as networks, cultural activities, confidence, connections, and commitment to shared local values. 'The social capital determines the capacity of a community to absorb shocks, exploit opportunities and direct it towards the future. Without social capital, community lacks cohesion, cannot organize to maintain its economy or its environment . . .' (Beeton 2006).

Methodology

The research was carried out over a three-month period. It entailed three exploratory visits to the village of Maâmar to establish relationships and plan the research to be done. The data collection was done while focus groups used tools to generate information on 1) the management, use and marketing of aromatic and medicinal plants and 2) organizations within Maâmar and the relationships among them. Focus groups were organized on the basis of gender (men and women) as well as of members and non-members of organizations in Maâmar. Gendered focus groups were used to avoid the influence of men over women as well as to differentiate perceptions by gender. Three SAS tools were chosen to meet these criteria.

The following table (10.1) shows the kind of information that each of the tools were chosen to generate and the numbers of men and women who participated in the focus groups.

The process used with each of the tools is described below. The force field and social analysis of collaboration tools were used with the same focus groups at different times.

Time line

We explained the purpose and process to all men and women. The older people in the groups instructed the younger members to draw. The participants indicated the

Table 10.1 Tools evaluated

Tools	*Why they were chosen*	*with who* *M*	*F*
Time line	To clarify the history of different organizations and their interaction within the village; to illicit information that might explain the relationships among them; to identify existing conflicts among organizations and between community members and the organizations.	21	17
Force field	To detect problems of different organizations, the causes of these problems, the factors permitting them to cope with these problems; the solutions that have been found.	14	11
Social analysis of collaboration: conflict, legitimacy, interest, power (CLIP)	To determine power relations, issues of interest and legitimacy, thus set up the profile of actors that affect the level of social capital in Maâmar. This analysis also determines the relations of cooperation and conflict existing between the various actors.	*	*
TOTAL		35	28

* Force field analysis and CLIP were done with the same groups at different times

key events and their dates. These were put on cards, assigning a title to events agreed upon by all. Then the cards were placed along the line in chronological order and stories were told about each one.

Force field

A list of key problems was made for each organization in the village separately, using a brainstorming technique. After the group had ranked the problems, the purpose and process of the tool was explained. The most important problem for each of the organizations was written on a card and placed along a horizontal line. Factors perpetuating each problem were listed above the problem card and those mitigating it were listed below. Then the group assigned a value to the importance of each factor on a scale of 1 (very little) to 5 (great). The results were then discussed in relation to the degree of control that could be exerted by the community on the perpetuating factor.

Social analysis of collaboration: conflict, legitimacy, interest, power (CLIP)

We explained the concept of social capital to the participants and asked them to identify those actors that can influence the social capital of each organization in Maâmar (Jamaa, Farmer's Association, Cooperative, Beekeeping group and Medicinal and Aromatic Plant group). Then the participants determined which of

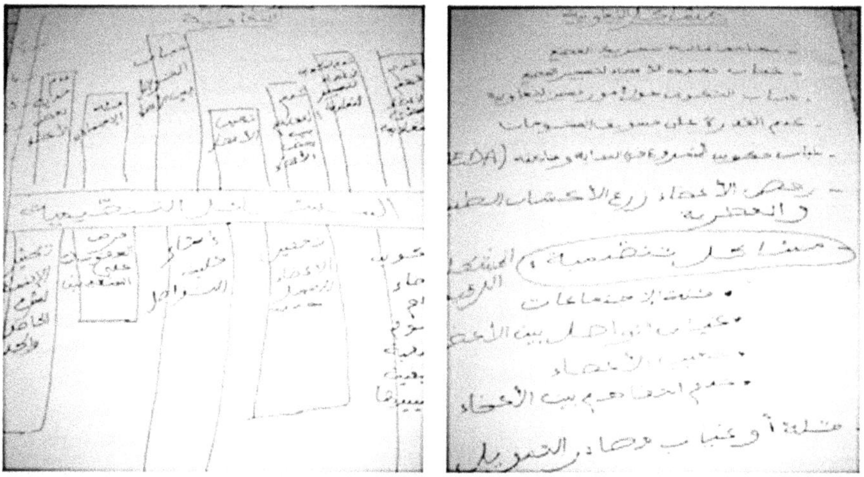

Figure 10.1 Force field analysis of the constraints facing the different organizations

the groups had high, medium and low levels of power. Subsequently the group provided their perceptions on the following questions:

- What does the cooperative do?
- Who makes the decisions in it?
- Who are the other organizations they work with?
- What kind of activities do they do with other organizations?

On the basis of this information, the participants identified the capacity of the organization to enhance social capital in the community. Then the group proceeded to identify and describe the degree of legitimacy using the same scale they had used to assess power. Once CLIP profiles had been prepared for each organization, a table was compiled summarizing the profiles and scores for all of the organizations in Maâmar.

Table 10.2 Organizations and their CLIP profiles

Category	*Description of the organization*	*Organization*
Powerful	Legitimacy + Power medium + Interest	JMAA
Dominant	Legitimacy + Power + Interest	Farmer's Association
Vulnerable	Medium legitimacy + Interest	Cooperative
Marginalized	Interest	PAM group
Marginalized	Interest	Beekeeping group

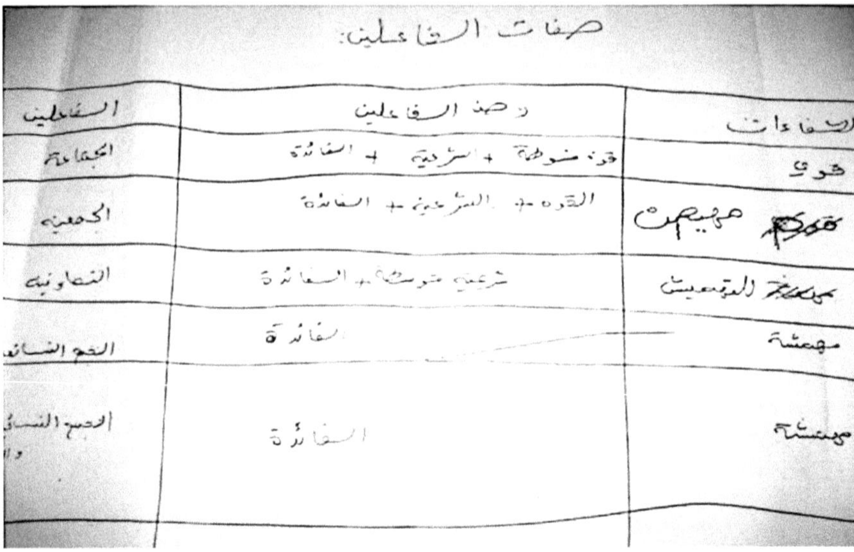

Figure 10.2 Social analysis CLIP on power relations, issues of interest and legitimacy

Analytical framework

The usefulness of the SAS tools was measured on the basis of their ability to illicit accurate information that can be used not only by researchers, but also to contribute to enhancing social capital within the community.

Criteria used for assessment

The objective was to contribute to an understanding of the usefulness of SAS tools from a gender perspective by comparing the comparative appreciation of women and men through 1) ease of control, 2) satisfaction with the information generated and, 3) its capacity to enhance social capital.

1 Ease of use and control of the tools

To assess the level of control over tools (comprehension and manipulation), we observed men and women using the tools. To measure comprehension, we used the number of reiterations the subjects required of us to understand the tool before using it. We observed the men and women during the sessions, identifying those who had difficulty (for example asked others for explanations). During the observation we noted the degree of difficulty each participant had using each of the tools. The comprehension and use of each of the tools differed by gender.

2 Satisfaction with the information generated

The tools were also evaluated according to the amount of satisfaction (good, medium, poor) participants felt with the information generated as a result of using them. Satisfaction was measured on the basis of two criteria:

- the perceived usefulness of the information for the future (for example to solve problems, to improve a situation);
- whether the experience of generating information made something new visible (for example a new perspective on their daily life, organizations in the community).

3 Contribution to the enhancement of social capital

The effects of the three tools on the construction of social capital were measured using the following criteria:

For individual social capital:
- increase in self-esteem;
- recognition for the individual from the group

Trust was built between:
- members and non-members and the organizations;
- members of the same organization and of different organizations;
- non-members (other people in the community).

For collective social capital:
- trust was established between organizations and with the wider community;
- cohesion of the group or organization was improved;
- new networks were created.

Analysis of the data

Descriptive and content analysis were used to understand the results

Content analysis was used to analyze the results of individual interviews. The entire interviews were transcribed and keywords consistent with our research topic to gain information needed to answer our research questions were identified. The analysis was thematic, organised in conjunction with the indicators described above.

Descriptive analysis based on frequencies allowed us to make graphic representations regarding the ease of use of each tool in relation to satisfaction with the information generated. The experiences of women and men were compared for each of the tools.

Major findings and discussion

The findings resulting from the assessment of the three SAS tools were analyzed by comparing the degrees of satisfaction that men as opposed to women have with each of the tools, taking into account:

- the ease of use and control of tools;
- the satisfaction with the information generated;
- the contribution to social capital (individual and collective).

The results were obtained by triangulating observations of the researcher and opinions of the participants in the research on aspects of social organization and gender in Maâmar.

Table 10.3 Major findings

Major finding *Control*	*Satisfaction*	*Social capital*
The control of tools to generate information on organizations was found to be, in general, easy. – Control of tools dealing with organizational issues is more difficult for women. – Women's difficulty with controlling tools that refer to organizational issues is consistent with the fact that they are not as involved in organizations as men.	– Tools that are difficult to control give satisfactory results to both women and men. – Satisfaction remains average for women because they do not know how they can use the information generated by using tools later in their lives. – Men are generally satisfied with the information dealing with social organization in the village, generated by using tools, despite the fact they found them difficult to control	– Tools that deal with organizational aspects and are difficult to control are those that bridge the two forms of social capital, individual and collective. – The social capital of women is not as strong as that of men. Using the tools helped strengthen women's social capital but in lower proportions than in the case of the men.

Control over the tools

The comprehension of the tools themselves and the information generated using them differed between men and women. The degree of control was assessed to be: very easy, easy and difficult. Figure 10.3 below shows the differences.

Control over all three tools was considered by women to be more difficult. This difference can be explained by the fact that the women tend to be mono-lingual, Amazhir speaking, have less education and wealth and lack experience with organizations in the community. Not surprisingly, and most likely due to their limited

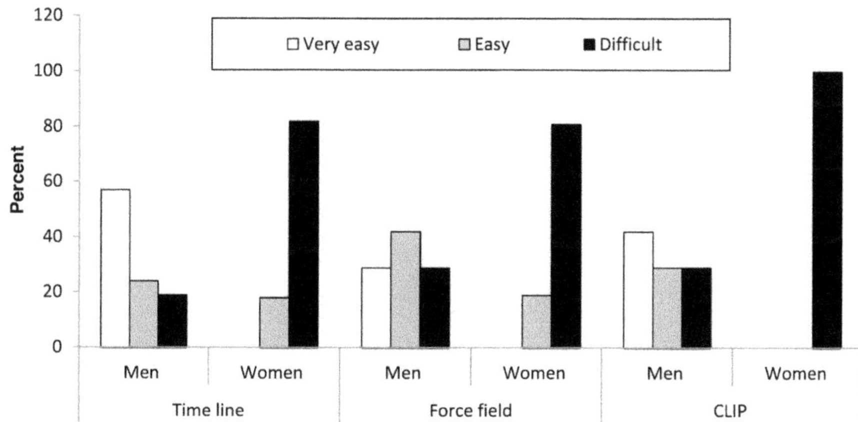

Figure 10.3 Ease of gaining control over the tools by gender

experience with organizations in the community, all women found the use of the tool that analyzes organizational dynamics such as conflict, legitimacy, interests and power (CLIP) most difficult.

Although new concepts and analysis resulting from the use of the tools were new to both men and women, men have more education and are active in community organizations, so they know a lot more about organizational history and dynamics than women.

The difficulty women have dealing with organizational analysis, while the majority of men found it an 'easy' to 'very easy' task, can be attributed to their relegation to participation in informal (AMP and Beekeeping) groups and their lack of involvement in the Cooperative and the Farmer's Association where strategic decision-making takes place.

Satisfaction with the information generated by a tool

Satisfaction with the information generated using the tools were also evaluated on three levels: high, medium and poor (although no one rated their satisfaction as

Table 10.4 Ease of using the tools by gender

Tools	*Degree of satisfaction*	
	Men	*Women*
Force field	High	Medium
Time line	High	Medium to high
Social analysis CLIP	High	Medium to high

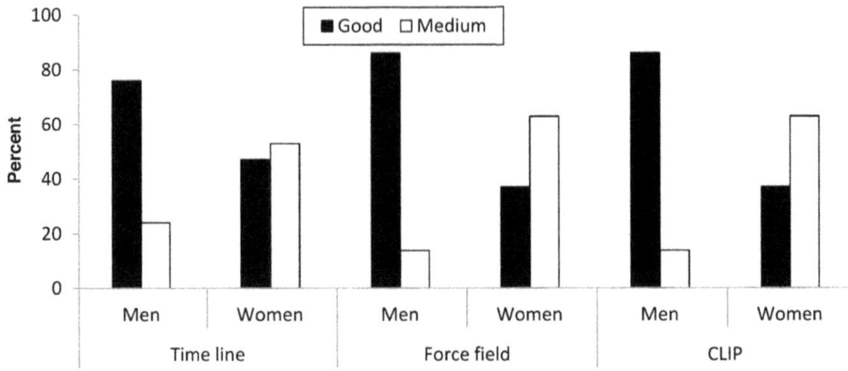

Figure 10.4 Satisfaction with the information generated by gender

poor). The graph above shows the degree of satisfaction of women compared with men regarding the information generated by each of the tools.

The lower satisfaction on the part of women can be attributed to the fact that they do not know how they can use the information generated for future benefit. In fact, they had difficulty identifying organizations in the community and the dynamics related to them (for example function, roles, goals). In spite of this, both women and men found the information interesting and felt that it provided new insight and clarified their knowledge of organizational dynamics in the community.

Mouna (woman, member of beekeeping group) said:

> 'The use of these tools allowed me to open my eyes to the truths that I did not notice before, because for us (women) all organizational issues are an exclusivity of men. Even for the women's organizations it is men who really decide... Now I know that we can play an important role in organizations... But I really do not see how I can use the information generated by these tools later for profit.'

Overall, it was those people used to being marginalized or to acting on the recommendations of others (women, the elderly, the poorer and the illiterate) who could not envision gains from participating in this kind of dynamic, action-oriented research.

The capacity of the tools to generate information to explore, analyze and better understand social relations within community organizations was greatly

Table 10.5 Degree of satisfaction with the information generated

Tools	*Degree of satisfaction*	
	Men	*Women*
Force field	High	Medium
Time line	High	Medium to high
Social analysis CLIP	High	Medium to high

appreciated by the men who are active and at the heart of this dynamic. They found the information to be of great value. On the other hand, the tools allowed women to make a self-assessment of the groups that exist within the village (SELF, Beekeeping, Cooperative, Farmer's Association, AMP cooperatives and Jmaa). The exercise allowed them to better understand the organizational elements within the female space of Maâmar and their own perceptions of the groups that some of them are members of. The opportunity to criticize, discuss and interact together with other women has contributed to both individual and group empowerment.

Fatima (woman, non-member) explained:

> 'Now I feel that I know the important role of the groups in our village, including women groups. Maybe I'm not member of any group but I am capable of making decisions. I can give my opinion and others have to listen to me. What we did (Focus groups using SAS tools) has given me the idea to adhere on one of women groups…'

The effects of SAS tools on the individual social capital

In order to assess the effects of the tools on social capital (individual and collective) we asked women and men who participated in the different focus groups questions regarding the degree to which 1) their self-esteem was increased, 2) the individual felt recognized by the group and 3) if they felt there was trust among the participants.

One hundred per cent (100%) of women and men who participated in the focus groups where the tools were used affirmed that their self-esteem was increased by the experience. They stated that the tools allowed the participants to express themselves, share their views, learn new things, and tell stories about their organizations and their problems. They felt that the tools helped to find solutions to the issues they identified.

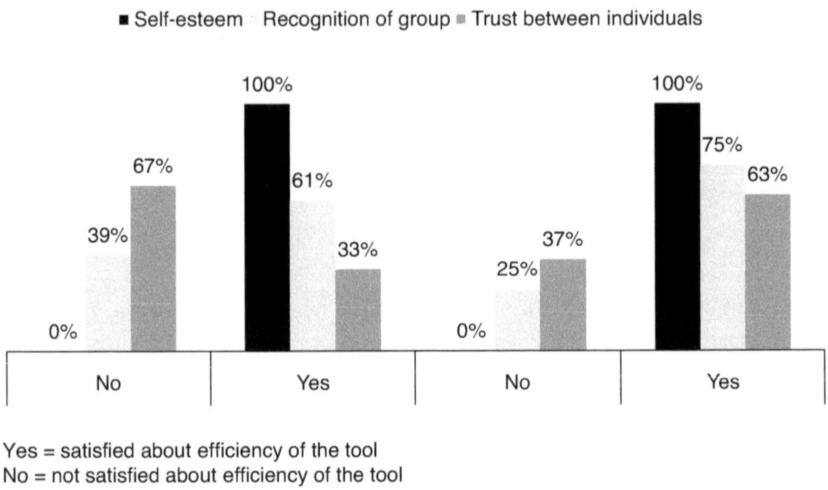

Figure 10.5 The effects of SAS tools on individual social capital

Lahcen (man, non-member) explains:

> 'Being in front of a pen and paper is already new for me and for others like me who have never attended school. To overcome the fear of taking the pen is a big step . . .' (This opinion was shared by some men as well as by the majority of women.)

Yamna (woman, member of PAM group) adds,

> 'It's not easy to give advice when you are just accustomed to follow without question. For the first time you feel integrated into the whole dynamics of organizations. For us it is something very important . . .'.

Khaddouj (woman, member of bee-keeping group) said:

> 'I do not know many things, but now I'm not afraid to speak because this allows me to feel the others' gratitude and appreciation for the little I do know and even for the courage to speak . . .'

Seventy-five percent (75%) of men and 61% of women stated that using the tools won them group recognition. They explained that expressing what 'we' think and giving 'our own' point of view is important and allows others to 'know you' better and appreciate 'what you know'. They added that 'not knowing everything' is not a problem, the real problem is to be afraid to speak.

On the other hand 25% of men and 39% of women explained that for a long time they had stayed away from the organizations in the village. They affirmed that it's not easy to be recognized when you don't know what is going on so don't have the courage to say anything.

Hassan (man, non-member) explained:

> 'Why would the others admire me or consider me more? It's true that our participation in focus groups and using tools allowed us to express ourselves and learn new things...But, I think, there are other persons who know more and who deserve to be recognized more than me.'

Sixty-three percent (63%) of men and 33% of women participants found that using the tools helped to establish trust. They explained that talking with others allowed them to understand things that for a long time had been ambiguous. This led them to overcome prejudices and trust others more.

Aicha (woman, non-member) added:

> 'To identify the problems of the different community organizations, understand them, study them and try to find solutions, helped me to understand the significant efforts of their members to overcome these problems. And so helped to establish mutual trust.'

These opinions led us to think that the major cause of the lack of trust among members of the organizations and with the rest of the community is a lack of communication and information sharing that can be overcome, at least initially, by making use of SAS tools.

Participants noted that the use of concepts like conflict and partnerships was new to them. They explained that their custom is to tend to stifle all conflict rather than explore the causes and project the consequences of it.

Hajj Bendihi (man, active member of the Farmer's Association) explained:

> 'Maybe there is conflict over the division of tasks...but it's not a real problem because those who refuse are sanctioned. The sanction is Izmaz: the person who opposes the opinion of the group has to prepare a meal for twenty or thirty guests and that is not easy to do. After this sanction the person will not try to impede the implementation of a collective decision.'

It seems that this self-repressive behavior has always been an obstacle to the expression of diverse points of view, resulting in a degree of mistrust among individuals. Another member explained that seeing the interest others had to better understand what is going on led the group to rethink the perceptions they had of other members of the community; for example that they are not interested in what happens within the organizations. He stated that the use of the tools led him to trust others more and to share more information. Others objected to these statements, explaining that it's normal that you cannot trust easily: 'for a long time there has been a lack of confidence due to a lack of information sharing and the marginalization of some people'.

In summary it would seem that people felt that transparency, effective participation of all of the community, and information and decision sharing precludes the building of trust. The use of the tools was instrumental in building enough social capital so that community members, whether they belonged to organizations or not, were able to analyze the causes of the lack of trust that came out as a key issue in the community.

The effects of SAS tools on the collective social capital:

The degree of social capital influenced by the use of the SAS tools included: 1) trust established between organizations and with the wider community, 2) improved cohesion within the group or organization and 3) the creation of new networks.

Fifty-nine per cent (59%) of women and 71% of men found that the SAS tools helped to build trust among organizations and between the organizations and the rest of the community. The tools helped the participants in the study to understand what hinders the development of the organizations in the community. They changed the generally negative perceptions they had of them and began to understand their logic.

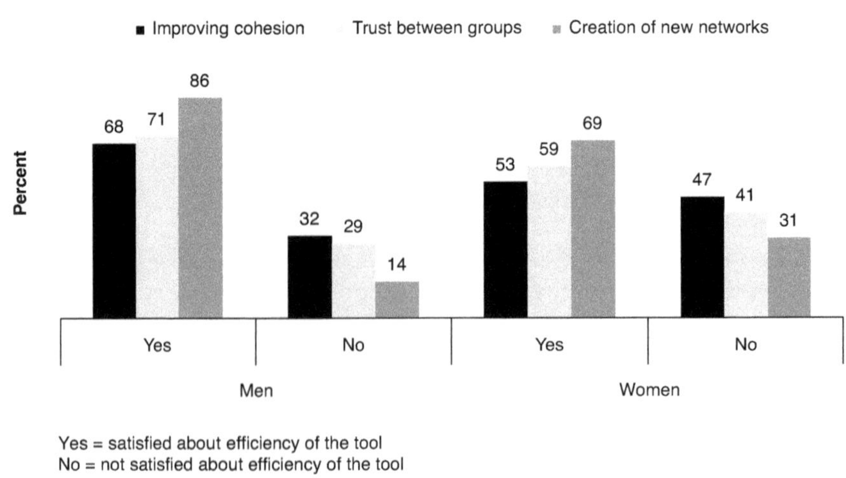

Figure 10.6 The effects of SAS tools on collective social capital

As a result, ground was laid for a change in the quality of relationships among the organizations and between them and the rest of the community.

> Upon completing the discussion of the force field Ahmed (man, non-member) told us:
>
> > 'Before I believed that these organizations, especially the cooperative, do nothing. But after participating in the discussions, I understood that the organizations had done many things to overcome their problems. What we discussed is a basis for establishing new relationships based on trust. I think it's our duty to pass what we have known to those who could not participate…'

Twenty-nine per cent (29%) of men and 41% of women who did not see that a basis for trust was being built among the different organizations argued that given the history of the problems, conflicts and partnerships, in and among the organizations in Maâmar, they cannot be trusted.

> Ittou (woman, non-member) explained:
>
> > 'For a long time, these organizations have ignored us, and they decided to work alone, their arguments do not convince me or many people in the village. If they want our trust… they have to work hard to earn it.'

Eighty-six per cent (86%) of men and 69% of women said that by using the SAS tools during the study, a new relationship among organizations and between them and the rest of the community had been created. Participants told us that it is necessary to build new relationships based on trust so as to bind the organizations together and bind them with the rest of the community. Even those who felt that trust does not yet exist said that new relational networks would allow for a better understanding of what is actually going on, and would eventually establish a solid foundation for respect and mutual trust.

> Ibo Mona (woman, member of Beekeeping and PAM groups) said,
>
> > 'I am member of the two informal groups. Before we did not talk about what is happening in the two groups with other women. But now there are women non-members who ask us about what is hindering our functioning… Even men of the Cooperative and the Farmer's Association asked us what we think about the problems… and I believe these are new links that result from what has been done…'

The 14% of men and 31% of women who thought that new networks were not fostered explained that despite the claims, the organizations have not been honest: 'ever since they were created..., it was the richest and those with more power (elected, appointed) who were chosen. They have ignored us for a long time and we do not believe that forging new relationships with them is possible.'

In the same vein Hamou explains:

> 'As a member of the community, the arguments given by the members of the organizations do not convince me to make a break with the past and forge new relationships easily. I think that the organizations must work more to gain our confidence...'

As for improvement in group cohesion, 68% of men and 53% of women said that the use of the SAS tools provided a basis for improvement. They explained that the work they did using the tools helped to establish a basis for mutual trust: by sharing ideas about how to improve the present situation and building support among the various parties. They felt that the research, using SAS tools, helped to improve the cohesion within the different groups and even among the various organizations by way of their members.

Moha explains (man, president of Farmer's Association):

> 'I am president of the Farmer's Association, and I will see that after the work done with these tools, we meet more to discuss issues that have emerged using them. These discussions have enabled us to share many things, with our members, with those of other organizations or with community people... This will help strengthen relations among the various parties and enhance group cohesion.'

On the other hand, 32% of men and 47% of women found that the work done with the tools was not sufficient to improve group cohesion. They stated that if real trust and new relations have not been created as yet, it is too soon to speak of improved group cohesion.

Ali (man, non-member) added:

> 'I do not think it's easy to improve the cohesion of groups. It is a very difficult job that requires honesty and seriousness. I do not think all organizations have this.'

He further explained:

> 'Organizations are judged by the character and behavior of their members; when some of these are dishonest, improving group cohesion will not serve their interests.'

Conclusion

As we bring the study to a conclusion, we would like to recall the objectives that have animated this project. We were interested in studying the effects of the social analysis systems (SAS) tools on building social capital among groups of women and men in rural communities. Indeed, the findings provide initial answers to the research questions using the criteria developed.

A gender analysis of the difficulty in controlling the three tools dealing with organizational aspects shows that women are distant from the organizational spaces in Maâmar despite their timid presence in the informal AMP and beekeeping groups. They remain structurally distanced from the strategic decisions of the Cooperative and the Farmer's Association. In spite of this disadvantage, it was found that women, as well as men, can gain by and improve their experiences by using SAS tools.

The capacity of the tools to generate information to explore, analyze and better understand social relations within community organizations was greatly appreciated by the men who are active and at the heart of this dynamic. They found the information to be of great value.

On the other hand, the tools allowed women to make a self-assessment of the groups that exist within the village (SELF, Beekeeping, Cooperative, Farmer's Association, AMP cooperatives and Jmâa). The exercise allowed them to better understand the organizational elements within the female space of Maâmar and their own perceptions of the groups that some of them are members of.

For the women, the opportunity to criticize, discuss and interact with other women has contributed to both individual and group empowerment.

The lack of transparency, dis-information and effective participation of the community has limited trust in Maâmar. The use of the SAS tools was instrumental in building enough social capital so that community members, whether they belonged to organizations or not, were able to analyze the causes of the lack of trust that is a key issue in the community.

A basis for establishing mutual trust, new relationships between organizations and with the community and even the sharing of ideas about how to improve the present situation can be attributed in considerable part to the use of the SAS tools.

So the research, using SAS tools, helped to establish a basis for improving the cohesion between different groups, and even among the various organizations through their members.

Since this study was carried out in a single rural community in the Atlas region of Morocco, replication of similar experiences in additional socio-economic, agroecological and ethnic conformations are in order. However, given our present findings, we are confident that the use of SAS^2 tools for gender responsive action-research can facilitate the construction of individual and social capital.

Notes

1 École Nationale Supérieure d'Agronomie (ENSA), Algiers.
2 Independent consultant, Peru.
3 International Center for Agricultural Research in the Dry Areas (ICARDA), Amman.

References

Chevalier, J. and Buckles, D. (2009) *SAS²: A Guide to Collaborative Inquiry and Social Engagement*. IDRC Edition ESKA, Paris.

Beeton, R. (2006) 'Society's Forms of Capital: A Framework for Renewing our Thinking' paper prepared for the 2006 Australian State of the Environment Committee, Department of the Environment and Heritage, Canberra.

Putnam, R. (1993b) 'The Prosperous Community: Social Capital and Public Life'. *American Prospect* 4(13): 35–42.

11 Exclusion of women from local development initiative doomed to failure

The case of the Middle Atlas community of Maâmar in Morocco

Fatima Zahid[1], *Siham Rohain*[1], *Karima Maâni*[1], *Sbai Mohamed*[1], *Maria Fernandez*[2] *and Aden Aw-Hassan*[3]

Introduction

Morocco's rural areas are characterized by poor socioeconomic infrastructure, low levels of education, inadequate support services and an ageing farm population. Farmers are thus not equipped to face the challenges of an economy that is opening up to free market competition. The major causes of insufficient productivity in the agricultural sector according to Akesbi (2006) are:

- production still hampered by climate constraints;
- human resources constrained by poverty and illiteracy;
- land structures unfavorable to modernization;
- farms and production systems still low intensity;
- insufficient and unevenly distributed financial resources;
- dangerously deteriorating natural resources.

Indeed, the annual cost of environmental degradation due to desertification, erosion, salinization, forest fires, loss of agricultural production and peri-urban farmland, is considerable. The latest cost estimate by the World Bank in 2000 would reach the equivalent of 4.59 % of GDP.[4] Another World Bank report adds that: 'the long-term sustainability beyond agricultural production has become more uncertain, because the crops have moved into marginal lands, deforestation has continued and industry tests the limits of water resources.'[5] The origin of natural resource degradation is often linked to the poverty of the population, too dependent on natural resources alone for survival. In fact, in Morocco, poverty is particularly high in rural areas, a situation that is favorable neither to the protection of natural resources nor to human resources development.

Poverty in Morocco remains essentially a rural and female phenomenon. According to the World Bank report on Poverty, published in September 2004, almost one in four Moroccans in rural areas is poor, compared with one in ten in urban areas. Although the rural population accounts for only 46% of the total population, rural areas host 66% of the poor, most of whom are women, and more vulnerable to environmental degradation. Indeed, women are often trapped in a vicious circle leading to degradation because they lack access to productive resources, institutional services, credit and technology. Thus, in order to survive, they are forced to overexploit land and forests that are already in the process of degradation. The mountainous regions are in particular danger given the level of natural resource degradation and unsustainable exploitation of these resources. However, there is wide recognition that consultation, participation, organization and gender are critical to the future development of these areas.

In the collective consciousness, as in the customary practices of civilizations, natural resources provide human beings with the necessary conditions for growth and existence. On the other hand, with the degradation of natural resources, the cultural heritage that determines their use gradually disappears. Expertise and non-standard knowledge of traditional systems in conjunction with the use of natural resources including aromatic and medicinal plants (AMP) become endangered. However, it is widely known that the rural populations, interviewed by social scientists, have repeatedly demonstrated extensive knowledge of their ecological environments (Dupré 1991 and Lévi-Strauss 1962). Men and women farmers still have extensive environmental knowledge drawn from their practical experience, historical events, observations, experiments and local cultures and traditions. This knowledge could contribute to the solution of current problems including deforestation, overgrazing and inappropriate farming practices.

Seeking to revive the agricultural sector in the face of an increasing global food crisis, the Moroccan government has decided to opt for a new strategy that focuses on improving the situation of farmers by creating employment and wealth. Indeed, Akesbi (2006: 119, 120) argued that an agricultural strategy must

> prepare farmers to be entrepreneurs through training and professional organization, meet the needs of producers and consumers and develop the rural areas. The professions in turn must be organized in effective and credible structures that take into account the implementation of development strategies across a sector or locality. The Moroccan Green Plan[6] provides farmers with technical and financial assistance to increase their production. It aims at promoting and guiding private investment, through targeted subsidies, focused on activities that make better use of the country's agricultural potential and the protection of fragile ecosystems.

In addition to the critical state of natural resources, Moroccan society has been gradually undergoing profound social changes. These changes have affected rural organization, the role of women in society and gender relations.[7]

The evolution of organizational forms in rural areas has been slower than in urban areas because they are rooted in the consciousness, values and the mentalities of the population. However gradually, as a result of internal and external dynamics: colonization, migration, media globalization, globalization and modernity, we are seeing changes in values, traditional organizations, power structures and gender relationships. Individuals and groups that were immersed in the tribe and family, protected by traditional forms of solidarity and circumscribed within a system of patriarchy are now facing changes that are often difficult to manage or understand. Development programs and projects with their technicians, money, models and approaches, force change on organizational configuration and dynamics of rural spaces. External agents have an impact on the balance and relationships between the rich and poor, between women and men.

All development initiatives have gender implications in terms of the expected contributions of men and women and the impacts of the interventions. More specifically when the initiative is addressing issues that affect women the most, the exclusion of women will certainly lead to the failure of the initiative. Overlooking the gender aspects of development initiatives is therefore a serious design flaw in a development project which can be avoided.

Development efforts by external agents need local capacities and support mainly from the local community. Understanding and utilizing the local community structure and institutions and involvement of both men and women, is very important for development initiatives to succeed.

Throughout this paper, we present a case where a failure of a project can be attributed to the exclusion of women or lack of strong and explicit inclusion of women in the initiative.

The objective of this paper is to look at the impact of one development program on women and their empowerment in a specific agroecological and socio-cultural situation. We will analyze the organizational changes experienced by the Middle Atlas village of Maâmar in the District of Khénifra that was the 'beneficiary' of the Euro-Mediterranean Partnership (MEDA) project MEsures D'Ajustement ou D'Accompagnement). We will look at the impact of the project on traditional social organization gender relations in Maâmar over the past eight years since the project began its activities.

The development context of Maâmar

The province of Khénifra, where the village of Maâmar is located, covers three major agro-ecological zones: sub-humid, semi-arid and arid and is organized into 35 rural communes with a total of 655 villages. Of the approximately 1.341 million hectares, 46% is under rainfed cereal cultivation while 42% is forest (Figures 11.1 and 11.2). The average annual rainfall is 180mm. The population uses forest resources to raise livestock, provide for medicinal needs, heat their homes and cook. It farms cereal crops for food and fodder and markets surpluses.

Rural Khénifra is one of the poorest regions in the Middle Atlas with a poverty rate of 28%, more than 35% of the population is considered to be vulnerable and the human development index is very low. Maâmar is made up of 140 households and is one of the richest areas in aromatic and medicinal plants (AMP). It has access to multiple water sources and is surrounded by forest. Traditional farming systems are still in use and 50 of the households live on subsistence agriculture and livestock.

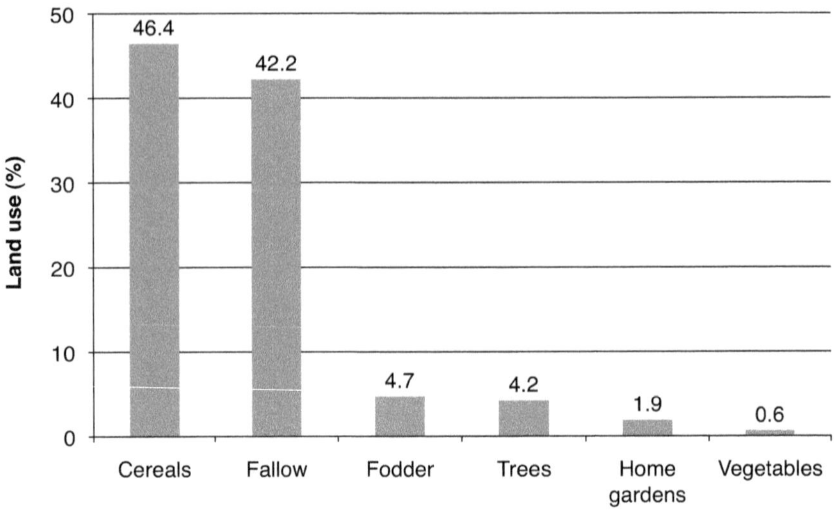

Figure 11.1 Cropping patterns in Khénifra (%)
Source: Provincial Direction of Agriculture of Khénifra 2009.

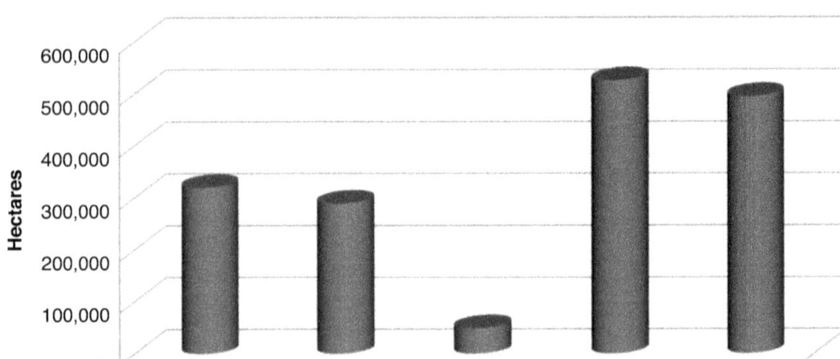

	Arable lands	Fallow	Irrigated	Forests	Rangelands
Hectares	320,000	289,250	49,000	526,000	495,000

Figure 11.2 Land use patterns in Khénifra (ha)

As a result, the intensification of AMP management seemed to have the potential for poverty alleviation by adding value and diversifying income generating activities in Maâmar. In order to reach this goal, the MEDA project (funded by the European Union) was implemented in Maâmar between 2003 and 2011. The project promoted the processing of essential oils to increase the value of non-timber forest products for medicinal, culinary and cosmetic uses. The project culminated in 2009 with the implementation of an AMP processing plant costing 200,000 US dollars. The five pillars of the project included: sustainable management of natural resources, reduction of erosion and drought damage, reduction of regional disparities and the promotion of the role of rural women.

The main outputs of the project were a mechanized AMP processing unit and producers' organizations to ensure efficient management of local processing and marketing.

The Association for the Territorial Management and Development (AGAT) of Maâmar supervises the Aït Bio cooperative responsible for the management of the plant. Both of these organizations were created in response to the MEDA project demand.[8] Although gender relations, organizational strategies and knowledge and skills required for the identification, processing and use of AMP existed in Maâmar before MEDA, it is not clear how much project authors and staff were aware of the role of gender for the success of the project. This gap will become clear later when we discuss the research problem in the context of this project.

Social organization of the community

Like other mountain villages in Morocco, Maâmar is organized around a complex system of written and oral Amazigh traditions that govern the sustainable management of natural resources, equitable distribution of investment in community infrastructure and conflict resolution. The families inhabiting the rural areas are grouped into villages (douars). The douars are regrouped into numerous tribes. Each village has a specific name, preserves its own historical and cultural identity, its boundaries and interests. The villages are most often administered by a village council (Jmaâ) that sustains both cultural and religious identity. The village council is responsible for implementing customary law (Azerf) that governs the creation and exchange of wealth in society. It operates according to the rules of Tiwizi: solidarity and cooperation in collective activities, one example of which is the system of collective caretaking (Tawala) and management of flocks of sheep on communal pastures.

In Maâmar, the council is composed of 13 members that include the elected Amghar (chief), the Moqadem, representatives of the two lineages of Maâmar 'Aït Boubrahim' and Aït Mhadji who are the managers and leaders of the mosque. Meetings of the Jmaâ take place after an incident requiring collective action; for example a conflict between the lineages or among those of the same lineage. 'If we have problems implementing a collective decision, the chief managers of the mosque, the Moqadems call a meeting to discuss the problem,' reports one of the managers of the mosque. The Moqadems communicate problems related to the management of the mosque, water and social life of the Amghar. The Jmaâ contracts the Fkih (see Figure 11.7) to manage the mosque. The Fkih announces and controls the five daily prayers, as well as the Friday preaching and religious celebrations. The contract stipulates the daily food and quantities of wheat, barley and olive oil he will be given.

The Jmaâ or village council has developed multiple solidarity strategies over time. Twiza provides material or physical assistance, free of charge, to any person in need. Community members organize Twiza to mobilize equipment and labor to assist widows, the sick and prisoners to plow, plant and/or harvest. In order to insure that a person in difficulty does not have to sell an animal at a low price, an Aouziâ is organized. Here a group of families purchase the animal (cattle, sheep or goat) at a fair price and then divide it among themselves upon slaughter. This system also allows community members to buy meat at a reasonable price without having to go to the weekly souk. Community members are involved in building the mosque.

Social solidarity under the auspices of the Jmaâ is demonstrated when a member of the community dies. Here, 'four members of the two lineages are elected according to their credibility and motivation to support the family of the deceased for three days.' (Lahcen, 60 years old). Water distribution systems that feed the traditional irrigation canals from a multiplicity of water sources vary between the two lineages. The Aït Mhadji distributes water by turns (Twala); once one farmer has used what he needs water becomes available to the next. The Aït Boubrahim on the other hand distributes water by a unit of time, usually a number of hours.

Inheritance rights to water are equal for men and women under Amazigh law. In fact, in Amazigh society women[9] have property rights to land and livestock and enjoy decision-making authority related to ownership and production that is protected by special statutes (Tamazallt). Customary laws (Izerfan[10] favor gender equality in the mountainous regions. Tribal representatives, in consultation with the foqahas,[11] ruled that divorcees and widows are entitled to half of the property of the marriage.

When a woman from Maâmar has a conflict with a person of the village around property or a resource (for example, inheritance), she asks the Jmaâ to solve it. In the presence of her family members, friends and neighbors, she invites the members of the Jmaâ to a meal. The Jmaâ discusses the problem and tries to solve it. If a compromise cannot be found, the women then recur to the tribunal at Aït Ishaq (district seat) or Khénifra (Yamna, president of beekeeping women's group).

The village of Maâmar has not shown increased social or economic development despite the investment and activities of the eight-year Khénifra MEDA rural development project.

Methodology

Between 2008 and 2011 the Development Engineering Department at the National Agricultural School (ENA) in Meknes decided to explore the reasons for the stagnation observed in Maâmar by way of a series of Masters level research projects. Three of these were carried out within the context of the ENA-ICARDA/SAGA project.[12] The first study focused on the contribution of AMP to household income in 2009. Over time, questions regarding women's role in the production and management of AMP and constraints on marketing of AMP emerged. Research concentrated on the knowledge and skills of women in processing and marketing of AMP. The study showed that women have knowledge, access to AMP resources and processing skills but do little marketing. While exploring these issues, it became clear that the attempts of the MEDA project to support AMP marketing had not been successful even though the community had mobilized behind the idea and considerable resources had been invested in a processing plant.[13]

The second phase of the research looked at constraints around marketing channels and found that the MEDA project had funded an AMP processing plant in the village to increase marketing opportunities. However, at the time of the study, the processing plant, managed by a dysfunctional cooperative, was idle. As a result the research focused on the impact of the MEDA project on gender and organizational relations in Maâmar. A research design was developed to look closely at traditional organizations and the gender relations within them and how these had been affected by the project. Key questions for the research included:

- to what extent could the failure of the AGAT (farmer association) and AIT BIO (cooperative) implanted by MEDA be attributed to the lack of women's participation in the governance and management of the intervention?
- why would a cooperative whose members were all men be managing an AMP plant when women are traditionally the producers, processors and marketers of AMP? Did MEDA take this fact into account when it formed the coop?
- how did AGAT and AIT BIO, formed under MEDA, affect existing village organizations and the gender dynamics within them? What is the cost of external development initiatives for local organizations?
- how were women's and men's rights over and benefits from the management of natural resources impacted?
- how has the apparent design flaw due to a lack of explicit and clear gender-sensitive processes which resulted de facto in women's exclusion affected the success of this development initiative?

Field research was carried out in four distinct time periods over a two-year stretch. The first phase of the research involved participant observation and the use of questionnaires that were administered to a total of 51 families. The respondents included 42 men and 23 women. During the second phase of the research participatory tools were used in gender disaggregated focus group discussions. This involved 66 men in eight sessions and 54 women in an additional eight sessions.

The following tables synthesize the various phases of the study, tools used, data collected and the number of participants by gender. The information collected was analyzed using thematic content analysis.[14] The focus group sessions were taped and analyzed using thematic analysis according to a categorical system.[15] This method allowed us to supply and justify answers with clear specifications. Benchmarking was also used to construct comparisons among different levels: gender, organizations and time periods (before, during and after) of MEDA.

Table 11.1 Timeline of research activities in Maâmar and participatory research (SAS) tools used for data collection

Date	*Activities*	*With whom?*
Feb. 2009	– Survey – Interviews	– 47 households – Responsible for the Provincial Direction of Agriculture, – Direction of water and forests direction – MEDA project staff – Herbalists and AMP companies
Sep. 2010	– First meeting with the field: networking – Determination of the first thematic	Members of organizations, random groups by gender
Oct. 2010	– Discuss the various problems that hinder its operation – Establishing relationships with these members	Community officials and members of the cooperative
Nov. 2010	– Development of resource maps and discussion around them – List organizations and institutions – Development ofVenn diagrams	Random groups by gender, some members of different organizations
Dec. 2010	– Mobility map – Venn diagram – Timeline – Seasonal calendar	– Women and men, youth, adults, old, rich and poor. – Member in organizations and those who are not. – In groups or individually
Jan. 2011	– Semi structured interviews – CLIP[a] (Collaboration, Conflict, Legitimacy, Interest, Power) – Map of organizational mobility – Matrix-conflict partnerships – Force field	

[a] A tool of SAS that helps create profiles of the parties involved in a core problem or proposed action based on four factors: 1) existing relations of collaboration and conflict 2) Legitimacy, 3) Interests and 4) Power.

Table 11.2 Participants in the study of social organization

Respondents to the interviews	*Lineage*	*Women*	*Men*	*All*
Villagers of Maâmar	Aït M'hadji	5	5	10
(not on village council or others)	Aït Boubrahim	5	5	10
Men's organizations (cooperative, farmer association and Jmaâ)		0	32	32
Women's organizations (beekeeping and AMP)		13	0	13
Total		23	42	65

Results

Gendered knowledge and skills to manage natural resources

Over the years, women and men have developed and inherited important local knowledge in the management and processing of natural resources, particularly AMP. Although a great deal of knowledge is shared between the sexes, the roles, activities and responsibilities of women in the management, processing and use of AMP is more specialized than that of men (Maâni 2010). It was found that both men and women could identify 12 species of medicinal and culinary importance (Figure 11.3). However, women identified more varieties of AMP than men (20% of women against 2% of men for lavender, 58% of women against 20% of men for flax-leaved (Daphne) pigweed and mint) while men identified slightly more varieties of marjoram, thorny capers and thyme than women.

Women and men all agree that knowledge related to the AMP has suffered severe erosion highlighting more extensive use of AMP in the past for cooking, medicines and cosmetics than today. It was found that the degradation of forest resources has led to some attempts at domestication. Figure 11.4 shows growing interest in domestication and expectation of utilizing medicinal plants as a source of income in the future. This trend may be partly the increasing number of initiatives on medicinal plants which raised expectations among rural people. It also compares the knowledge of AMP for specific uses by men and women.

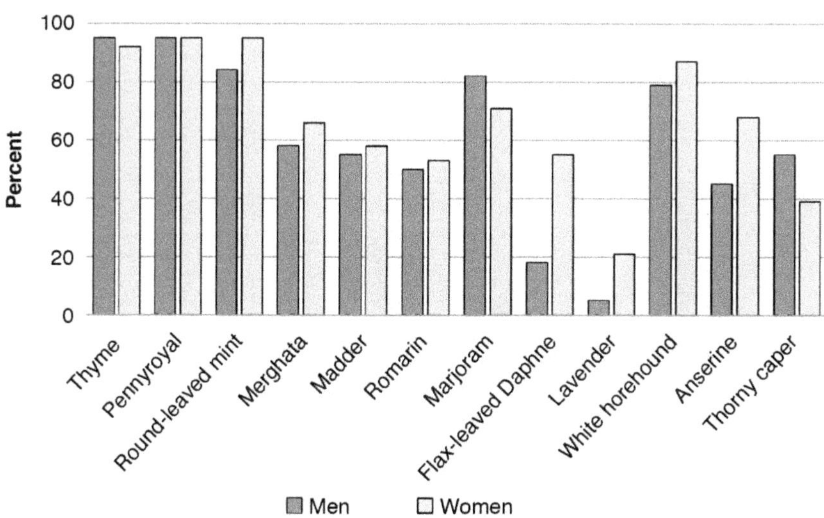

Figure 11.3 Knowledge of the main aromatic and medicinal plants by gender

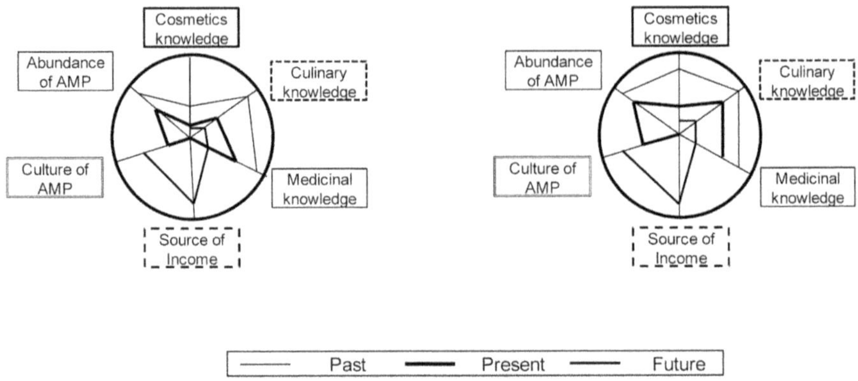

Figure 11.4 Local knowledge of men (left) and women (right) about aromatic and medicinal plants

Marketing strategies and income from AMP

The research conducted by Sbai in 2009 coincides with the end of the MEDA project. She showed (Figure 11.5) that at the beginning of the project, AMP was important in the household income. AMP contributed 8% of annual household

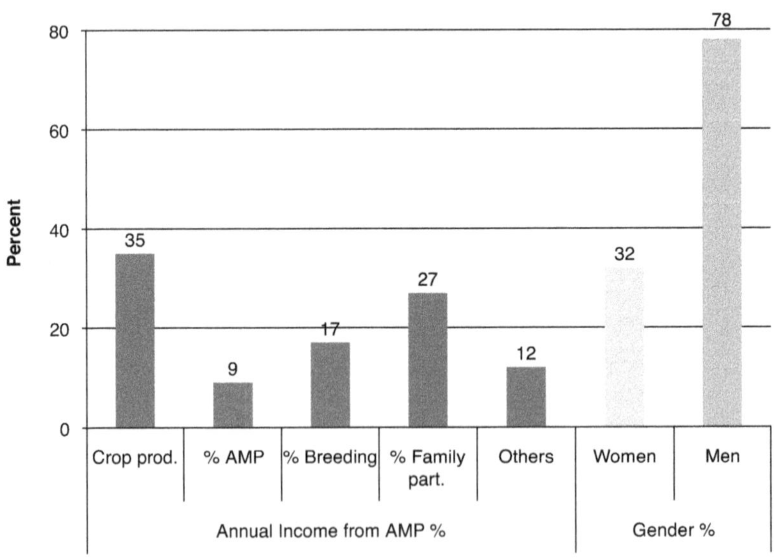

Figure 11.5 Annual income from AMP men and contribution of women and men to the annual income

income; 78% of this income could be attributed to the knowledge, skill and labor of women.

The contribution of AMP to family income in the years before MEDA project (Sbai 20009) can be explained by the fact that the project used its relations to identify spaces where traditionally processed AMP production (for example festivals, fairs) can be marketed.

Social organizations in Maâmar during the MEDA project

In Maâmar, social life has been managed by Jmaâ, by their ancestors and by all the rules they have inherited since the centuries. But this organizational life changed in 2003, the year that the Association of Territorial Planning and Management (AGAT[16]) was created by the MEDA project as a requirement for the village to be able to benefit from development projects in Khénifra. The AGAT is a form of organization recognized by the Moroccan government where the Jmaâ is not. The (male membership of) AGAT can make requests and negotiate with government authorities under the supervision of Jmaâ. But this was just the beginning; more organizations would be formed within the context of MEDA: a formal cooperative of men (Aït Bio in 2007) and two informal groups of women (2004, 2007).

Social organizations and women in Mamaâr after the MEDA project

At the time of this study (2008), there were two formal and three informal organizations in the village: Jmaâ (village council), AGAT (Farmer association), Aït Bio (cooperative), women's beekeeping group and the SELF-AMP women's group. Only AGAT and Aït Bio are recognized formally. Intermediate findings during the research pointed to the fact that the organizational structures created during MEDA may have had a negative impact on social capital and especially on gender relations within Maâmar since 2003. In order to understand the impact better, an analysis of these structures was done in the light of the motivations behind their creation; and the nature of links and relationships among them (for example leadership, survival and power). Table 11.3 provides an overview of the organizations: how they came about, their status, the purpose and degree of overlap among their membership.

AGAT: the state-recognized farmers' organization

AGAT was created in 2003 as a result of the MEDA team's influence on the community members who were interested in benefiting from the project. A village committee composed of members of the Jmaâ and some leading members (sons of Jmaâ members)[17] chose the membership respecting the two-lineage composition to prevent future conflicts. AGAT provided the village with a formally recognized interlocutor between the villagers and the project as well as with government agencies. The village committee met to elect the officers of AGAT. At the beginning, AGAT had nine members, most of whom were illiterate.

Table 11.3 Organizations in Maâmar after MEDA

Organization	*Status*	*Created*	*Who formed it*	*Purpose*	*Members*	*Degree of overlap of members*
Jmaâ	Informal	Unknown	Population	– Personal conflict management (inheritance, divorce, etc.) – Management of the mosque – Management of water towers, sources – Management of conflicts around fields and forest access – Penalties to village members for various reasons	13 men	1 member of Jmaâ is also in AGAT
AGAT	Formal	2003	Jmaâ on insistence of MEDA project	– Planning trails and irrigation ditches – Organization of the population around the MEDA project – Distribution of plants, apiaries and equipment – Organization of training with the MEDA project	11 men	5 members of AGAT are in the Aït Bio
Aït Bio	Formal	2007	Jmaâ on insistence of MEDA project and AGAT	– Recovery and transformation of AMP – Management and operation of the unit of distillation – Management of AMP marketing – External relationships with partners (other structures . . .) – Identification of and linking with markets	16 men	2 husbands and 2 sons of AMP women's group
Beekeeping women's group	Informal	2004	Jmaâ on request of MEDA project AGAT	Higher incomes for women	7 women	4 women are in the both women's groups
AMP women's group	Informal	2007	Jmaâ on request of MEDA project AGAT, Aït Bio	– Sowing, planting, gathering, drying the crop AMP – Collect spontaneous AMP – Higher incomes for women	10 women	3 wives of Aït Bio 2 wives of AGAT leadership

Later the group recognized the need for educated members to support project activities. One member stated: 'After the creation of the AGAT, we realized the need to integrate educated members to write reports, contact people outside of the village and to make visible what we do'.[18] The problem was solved through the integration of new educated members. The profile of the major historical events experienced by the AGAT is summarized in Table 11.4:

Women's beekeeping group

In 2004, MEDA proposed that AGAT form a group of women to benefit from an apiary project that was benefiting other villages in the Khénifra region. MEDA's objective was to create an income generating activity for rural women by building skills in beekeeping techniques. A list of 204 interested women was drawn up in an AGAT organized meeting. Only seven women were selected to form the group on

Table 11.4 Activities of AGAT

Date	*Activities*
2003	Creation of AGAT
2004	– Working of water source 'Tekechout'.[a] – Distribution of fruit trees (olive, almond, carob, caper and quince) given to the public at a price of a dh/plant for 4 years – Organization of the population to benefit from project activities – Distribution of the apiaries to women's beekeeping group
2004/2005	– Improvements to seguias of Maâmar. Population (all 140 families) participated with a small amount of money (50 dh) and provided labor – Additional funds and materials were provided by MEDA – Construction of a water source 'Maâmar', was funded by MEDA
2005/2006	– Construction of a water source to 'Bni Khlil' – Construction of a well 'Bousbou' – Operations were financed by the project; the land was purchased by AGAT
2006/2007	– Construction of a water source 'Ahfour Iksasen' 2km from Maâmar was fully funded by the project – Planning of a road from Maâmar to Bni Khlil (12km), there was also the construction of the fence of the Maâmar school; – Construction of the AMP distillation unit. The land for the plant was bought by AGAT and construction was funded by MEDA – Water and electricity are paid for by the cooperative and AGAT
2008	– Construction of toilets for many of the Maâmar households – Distribution of dryers for drying of medicinal and aromatic plants for women members of the SELF group – Training of women (sowing, planting, gathering and technical drying): 3 sessions, 2 days/session

[a] Amazigh name.

the basis of four criteria: motivation, community spirit, social status and previous experience in the field of beekeeping. The women in the beekeeping group are all from the Aït Boubrahim lineage.

Village women who are not members of the apiary group believe the constitution of the group was neither participatory nor transparent. They feel that only group members – and not the village as a whole benefit. The interviewed men mostly ignore the existence of the group or see it as irrelevant.

The Aït Bio cooperative

MEDA required the creation of a for-profit structure in Maâmar as a prerequisite for the donation of the AMP processing plant. MEDA considered that a cooperative would stimulate better organization, provide income and promote business opportunities for enterprises such as nurseries and distilleries for herbalists and farmers specialized in AMP. In 2007 Aït Bio was formed and the AMP distillation unit was installed. To form the cooperative, a meeting of men interested in AMP was organized by MEDA project personnel and the officers of AGAT appointed the cooperative officers. At the time of this research, the cooperative had 16 members representing the two lineages although the decision-making power was concentrated in the hands of only four as can be seen in Figure 11.6.

Interviews with villagers indicated that those who are not members of the cooperative ignore or are unaware of this unfamiliar structure and do not see the point of its existence. They felt that the creation process was unclear and that there was a lack of information on the procedures used for selecting members. They also considered that only a few villagers monopolize the decisions within the cooperative.

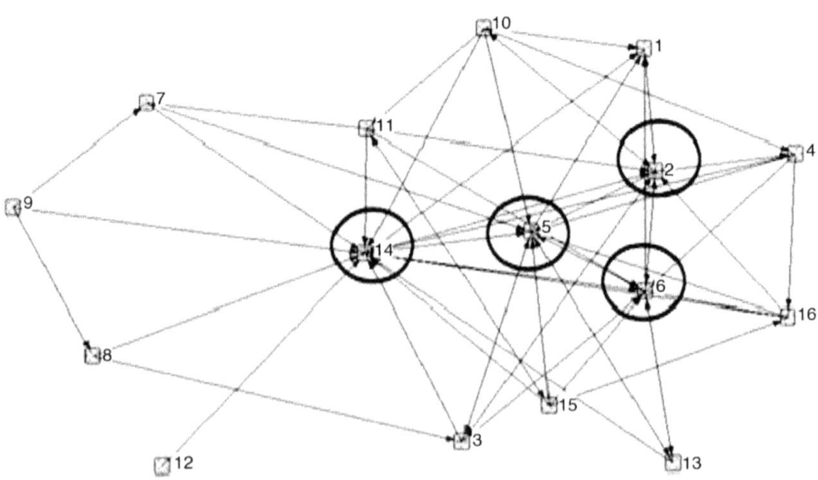

Figure 11.6 The communication network of the cooperative members (sociogram)[19]

SELF: women's AMP group

The MEDA Trust, AGAT and Aït Bio initiated SELF in 2007 after the installation of the distillation plant. MEDA's stated purpose for the SELF group 'was to further develop the role of women in the management, collection and drying of medicinal and aromatic plants to improve product diversification and household income'.[20] The three male organizations decided that the group would be made up of 10 women: seven from the beekeeping group and three from the Aït Mhadji lineage, leaving the majority with Aït Boubrahim. The group, with no formal recognition, would assist the Aït Bio members (all men) in the technical management of AMP by sharing their experience of domestication, collection and drying of wild species in the forests surrounding Maâmar. The SELF group would also provide raw material for processing in the plant. The members of SELF (women) and of Aït Bio and AGAT (men), is a continuum of the relationships in the domestic space as is shown below in Table 11.5.

Perceptions regarding SELF among the villagers vary. Men do not see the purpose of the group or its role in the life of the village. Women however do recognize the group but feel that it is only beneficial to its members and not to the community as a whole.

Social organizations and lineages in Maâmar

The plurality of organizational forms and structure in Maâmar leads us to question the power of lineages and the role of this variable in building influence and making decisions in the daily life of the community.

The Jmaâ is an assembly of 13 elected or designated men. The membership is divided equally between the two lineages. It is chaired by the Amghar (Chair), who is assisted by two Moqadems elected annually by their lineages, and the contracted managers of the mosques.

The majority of the AGAT's membership is from the Aït Boubrahim lineage (54.5%) while those belonging to Aït Mahdji lineage make up 45.5%. Members of the Aït Boubrahim lineage make up more than two-thirds of the cooperative membership. Although the two lineages are strongly linked by blood relationships, marriage and residential proximity, the Aït Boubrahim lineage is better represented in both formal and informal organizations whether by men or women. This gives the lineage more power in Maâmar (Table 11.5).

Table 11.5 describes the concentration of decision-making power within these organizations among three families whose members (whether men or women) are present in all of them as well as in the municipal council at Zaouit Aït Ishaq (district headquarters). Figure 11.8 gives examples of the relations between men and women in the organizations in Maâmar. The organizational mobility map is used to refine the organizational analysis of relationships of blood and marriage among the members of the different organizations. It also allows us to study the mobility of members among the different organizations and detect leaders, key people and the dominant lineages.

Table 11.5 Families and power in Maâmar

Number of women affected in the sociogram	*Group*	*Links of women to AGAT and Aït Bio members*
5	SELF	Wife of vice president of the AGAT Mother of one member of AGAT one of Aït Bio
2	Bee	Wives of Aït Bio members
1	Bee	Wife of AGAT president, sister of secretary general of Aït Bio and of vice secretary of AGAT
10	SELF	Wives of Aït Bio members
6	SELF	Wife of the vice president of Aït Bio

The Venn diagram (Figure 11.7) has allowed us to analyze the various perceptions of women and men about roles played by different organizations within Maâmar. It allowed us to identify the stakeholders that influence decision making in the douar. The Jmaâ and the AGAT are the most influential actors. All other organizations have no potential interest to the population. This figure allowed us to see if the population in Maâmar is satisfied or not with their actions, to identify also potential conflicts between different groups. It presents also an opportunity to discuss possible changes (greater involvement of particular actors).

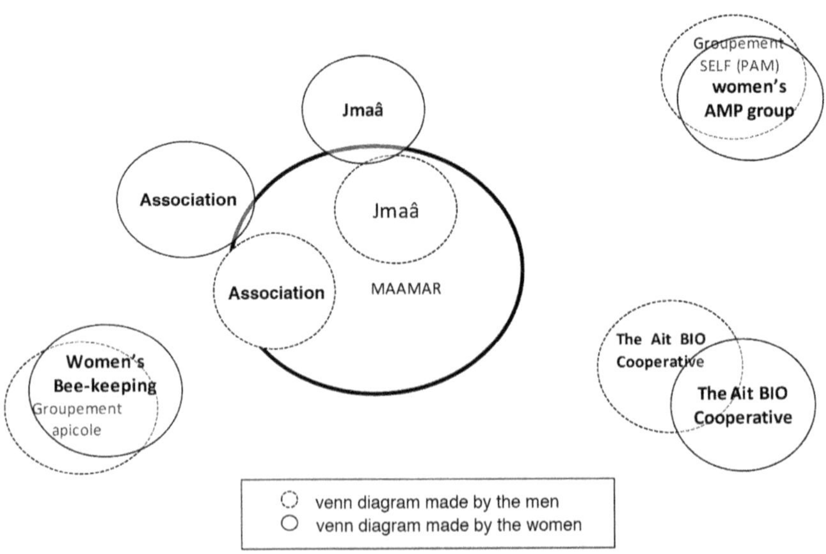

Figure 11.7 Importance and interaction among organizations by gender (Venn diagram)

Figure 11.8 shows the different family links that exist within organizations in Maâmar. It is used to identify the major families that share the organizations and the family traceability encountered in this space.

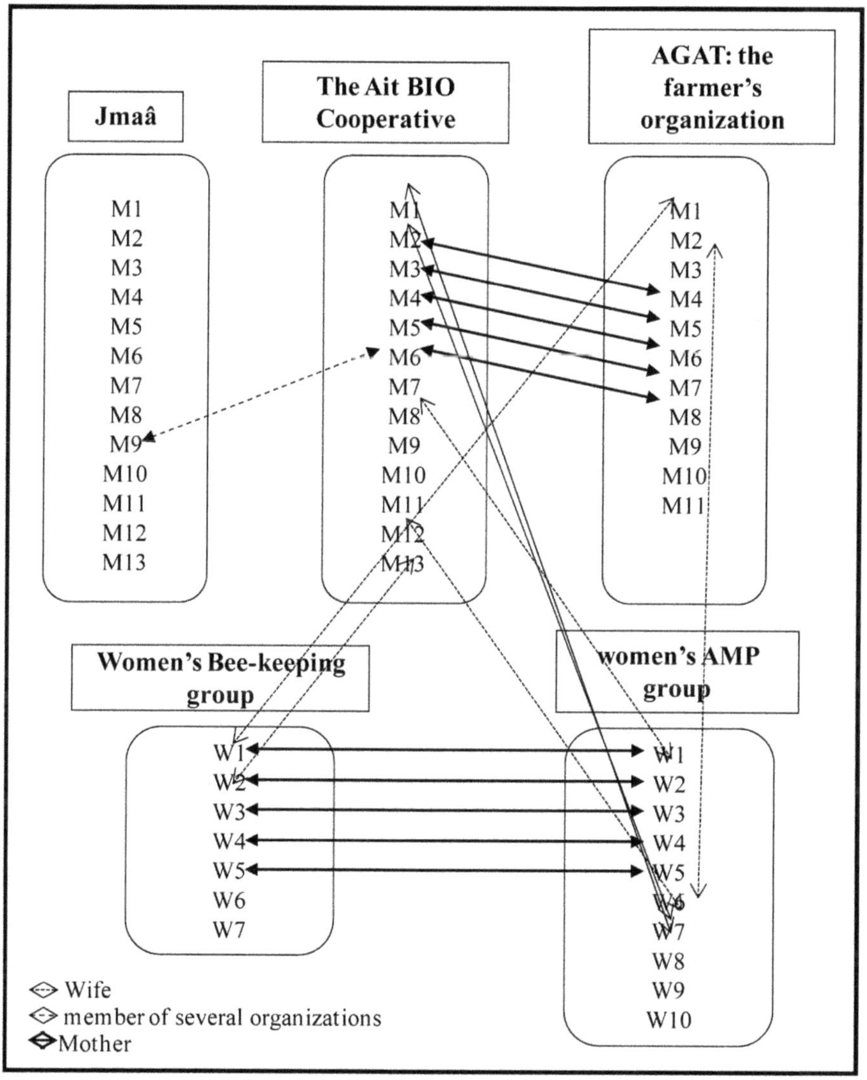

Figure 11.8 Relationship between members of different organizations in Maâmar (organizational mobility map)

Impacts of MEDA on social organization and gender relations

AGAT is the parent of Aït Bio and there is strong collaboration between them. They often meet together and coop decisions are negotiated. The cooperative maintains strong relationships with SELF because it depends heavily on the contribution of the women for raw material as well as for knowledge and skills. As it turns out, the women of SELF are responsible for collecting and drying the AMP while the men manage the processing plant and the marketing via the cooperative. This distribution of tasks by gender is not fortuitous. It reflects the relationship between women and men and demonstrates that although women hold the technological knowledge and skills for management of AMP, MEDA has awarded men the control over the plant machinery and marketing channels.

Our data leads us to focus on three impacts that the modifications in the organizational structure of Maâmar, resulting from the MEDA intervention, have had on the community.

The first impact

AGAT is an organizational modality promoted by the MEDA project, superimposed on the Jmaâ. In the mind of the project, this 'modern' organization would be better able to link the population to government agencies so that local interests could be better defended. An AGAT was formed in the village of Maâmar in 2003. The project determined the number of members the AGAT should have; the Jmaâ appointed 16 men following its own representative structure. Our findings show that the Jmaâ agreed to form an AGAT so that Maâmar would be able to benefit from project activities (irrigation canals, school and sanitary facilities).

Based on a diagnosis of natural resources in Maâmar, the income-generating component of MEDA decided to support the production, processing and marketing of aromatic and medicinal plants (AMP). However, the diagnosis did not take into account access to, control of or skills related to AMP by gender (Maâni 2010). This study shows also that women not only have access to the forest but that they spend a lot of time managing it. The forests of Maâmar (Boutoure, Bouhalfa Tissedelt) are the main sources of aromatic and medicinal species. Both men and women harvest AMP when collecting wood, but women perform the remaining management activities. Women can easily identify the agro-ecological habitats of each species in the forest as well as in the fallow lands and free places of the village. The plan developed by the MEDA and accepted by the AGAT (only male) has shifted customary responsibilities and recognized roles of managing and processing of AMP from women to men.

The second impact

The project argued that if AMP by-products could be processed using modern technology, these could be sold to export markets since there was a demand.[21] AGAT accepted the proposal but no feasibility study was done on the capacity of production of AMP in the village. It can be imagined that discussion within the AGAT regarding AMP yields in relation to the capacity of the processing plant was not based on experience since men do not manage AMP resources. As a result, the project built a 300-liter processing plant in the village when the production capacity of AMP raw material is less than 1/10 of what the plant requires to function efficiently. As a result, the all-male cooperative is left holding a white elephant.

The third impact

In order to receive the donation of the processing plant, in 2007 the project required the Maâmar AGAT to form a for-profit cooperative. The all-male membership of the cooperative was chosen by AGAT in consultation with the Jmaâ. Again, the cooperative reproduces the lineage and kinship and power relations that bind AGAT. Most meetings are held jointly and the decisions of the cooperative are often negotiated with the AGAT that interfaces with the Jmaâ. The concentration of the power in the AGAT is weakening the representation of the Jmaâ and eroding its authority.

The development issues AGAT works on are different from those that concern the Jmaâ and give it a new role in Maâmar society. Organizational and management skills developed in the context of MEDA have allowed AGAT to gain strength and compete with the Jmaâ. Even though the Jmaâ has opened itself to additional forms of organization in Maâmar (Aït Bio, women's informal groups) AGAT now provides legitimacy and credibility to the Jmaâ. This erosion of the Jmaâ's power is making this traditional institution less representative while making AGAT stronger.

The consequences of MEDA for the social capital of Maâmar

An analysis of the roles and functions of community organizations in relation to power was made and showed that the Jmaâ continues to manage conflicts in Maâmar. It has embodied the roots, history and tradition of Maâmar for as long as it has been a recognized rural community. These two characteristics provide greater legitimacy and power to the Jmaâ. The farmer association (AGAT) is the recognized and obedient son of the Jmaâ and so enjoys power in decision-making regarding the organizations that have been created after it. The legitimacy of the AGAT in Maâmar is high because the state recognizes it as the entity that can engage in legal contact with external organizations.

There is close cooperation between the Jmaâ and AGAT in decision-making. Indeed, the Jmaâ validates decisions taken at meetings by members of the AGAT. The population sees the cooperative as a creation of the AGAT. Although the cooperative is also a formal organization, it does not enjoy legitimacy among the population and has no power. Failure of the cooperative is seen as a failure of the AGAT. The women's groups (SELF and beekeeping) on the other hand are not only informal, they are not even recognized at the village level and have neither authority nor legitimacy. Relations between the cooperative and SELF are strong only because the knowledge, skills and labor of the women are critical to the production and processing of AMP that is now being managed by the cooperative.

A patriarchal power relationship continues to bind AGAT and other organizations created under the pressure of MEDA, to the Jmaâ. Only the Jmaâ and AGAT enjoy substantial power and influence in the life of Maâmar although the latter is less representative. The relationships that bind the Jmaâ to AGAT are similar to those that bind AGAT to Aït Bio and the latter to SELF. The relationships between organizations are hierarchical, tinged with a feeling of paternalism which links existing organizations and newly created ones. Being at the origin of its creation provides a sense of paternalism to the old organization which expects obedience from the new one. This suffocates the cooperative and the SELF in their decision making. The only group that has escaped this type of relationship is the women's beekeeping group that is informal with no direct links to any of the other organizations and so enjoys greater flexibility and autonomy.

It would seem that MEDA instigated the formation of new organizations in Maâmar disregarding the history, legitimacy and functions of the Jmaâ. Although the Jmaâ consented to the formation of the AGAT it had no choice given that it was a pre-condition for the village's participation in MEDA. Regardless of gender bias, the negative effect of the newly created organizations on the cohesion and capacity of existing organizations to adjust themselves to new opportunities is clear. Although the dynamics have resulted in the AGAT having some power, legitimacy in Maâmar is still maintained by the Jmaâ despite its informal status vis-à-vis the state. However, the division of power and recognition are resulting in a weakened capacity of Maâmar to organize and negotiate development.

It could be argued however, that the legitimacy of AGAT may be affected by the current lack of development activities and over time. Although AGAT is influential, the Aït Bio cooperative remains vulnerable and the two women's groups are marginalized.

The consequences of the MEDA project for gender relations

What is the role that women play in the new and old organizational configurations? Did the new organizations change gender dynamics in the community? Did it promote or depress the role of women?

The power relations between and among the different organizations in Maâmar are only one layer of the cake, if be the most visible one. The second layer has to do with the impact of MEDA on gender relations within the village organizational structure situated historically in the Jmaâ. Although MEDA had an idea that women should be 'on the map', it never actually was able to build a platform for conversation with them. In fact, it never requested the Jmaâ to bring women to the table but rather only looked for them after the all-male AGAT had been formed.

Although women are not visibly represented in the Jmaâ, the Jmaâ does have channels whereby women can make their voices heard and their needs attended to. This is not so in the case of AGAT. Women have been totally excluded from membership in the formal organizations created at the request of MEDA (AGAT and Aït Bio). The women's beekeeping group was fostered as a side activity and the

SELF AMP group was created to serve as knowledgeable and critical for the male cooperative. Both are informal.

In the end it is the men of the cooperative who are responsible for maintaining relationships with agencies and technicians, for maintaining the machinery and marketing the products while women provide the raw material and control the majority of management skill and technical knowledge about AMP. To make matters worse, the women do not seem to be leaders in their own right, but only as they are linked through marriage or family relationships to the leadership of AGAT and/or Aït Bio. In fact, any power that these women might have has to be validated by their male counterparts in AGAT and Aït Bio. Thus, in this Amazigh area, the power traditionally enjoyed by women inside their homes and in the community has been seriously altered and eroded as a result of the practices and organizations fostered by the MEDA project. The organizational dynamics imposed by the MEDA project has resulted in a decline in Maâmar's social capital. This has influenced livelihoods, the marketing of AMP and gender relations in this space.

What have women lost in this experience?

Women in Amazigh society and within the context of the Jmaâ have a right to property, natural resources and access to channels of defense and negotiation of their rights. However, the new formal institutions created by the MEDA project have changed the culturally semi-egalitarian relations into discriminatory ones. Women have been relegated to informal structures leaving them with less power than they enjoyed prior to MEDA. The introduction of women to the organizational processes caused them to lose the kind of power they were assured by the customary laws enforced by the Jmaâ.

Even if women have not lost power in their homes or with Jmaâ, they have lost the instantaneous force and channels by which to respond to actions and decisions that could affect them negatively. This experience affected their collective ability to meet, discuss and organize themselves because the failure of this experience has also negatively impacted on their individual capacity to take the initiative, to share it and develop it.

These gender relationships shared between the formality and informality of the structures are easy to see when we analyze the beekeeping group. It has successfully managed its business where SELF has failed. This success can be explained by the fact that beekeeping is not linked in any way to the AMP chain. It is discrete from male intervention and the concerns of the cooperative and AGAT. The increased empowerment of this group for decision-making demonstrates how women can succeed in the absence of discriminating power relations with men.

The members of a female organization maintain more intense relations, more discussions, and more communication among themselves. This reinforces the position of women within the organization, increases its power and results in greater managerial capacity.

The women of SELF are not full members of the cooperative and are relegated to perceptively lower value or less prestigious while more laborious activities, while men have taken over responsibility for equipment and marketing which have the opposite attributes. This leaves women with no decision-making power and without opportunities to develop new skills. Experience has shown that mixing men and women in the same organization does not encourage their development. The study by Tidaoui (2011) showed that in rural Morocco, women become more powerful in purely female organizations and when organizations are mixed, women lose power. Indeed, in mixed organizations women are confined to traditional roles and activities that do not allow them to gain power. However, when an organization is strictly female more women develop their knowledge, managerial skills and their negotiating power.

Implications for development initiatives

The multiple organizational forms fostered during the MEDA project have recreated and reinforced male-centered, hierarchical and paternalistic Jmaâ-type configurations and have relegated the participation of women to 'informal' spaces that afford no public voice. Although the government does not recognize the Jmaâ, it is recognized by all of the village residents as its legitimate representative body, confirmed by the fact that the post 2003 organizations fostered by MEDA must consult and report to it.

Although the Jmaâ retains power, it has been weakened by the creation of satellite and dependent organizations that limit its capacity to reinvent itself; for example adopting an assembly style and/or including women representatives. Although women can't presently become members of the Jmaâ, there is a mechanism by which they can make a claim when their rights are compromised. On the other hand, in the case of AGAT, women not only have no direct representation, there are no mechanisms by which they can claim their dues.

Organizations created by males that bring women in later ensure that women will be at a disadvantage as the power of men is enhanced. All women's organizations, however, can give them more power, increase their capacities and foster financial independence.

This research poses several questions regarding the impact of development projects: their social and technical feasibility, and their impact on the social life of rural people as well as on the relations between men and women. To what extent do the development policies of international and national organizations address the core economic and social equity needs of rural populations? Considering the persistence of poverty and insecurity in regions exemplified by Maâmar, it is essential to conduct a review of approaches and learn from existing development practices.

We know that a successful community development project ensures that men and women of a community decide collectively what their priorities are. However, it seems we do not know how to understand and analyze the social and organizational processes in communities, respecting their world views and their own roads to development. An important aspect of the latter is an understanding of the gender relationships in place so as to help them evolve positively, safeguarding and enhancing recognition of women's roles and their rights.

Conclusion

Women in Maâmar occupy an important place expressed through the traditional organizations of douars. In parallel with this village structure, other more modern forms (cooperatives, associations, groups) were born in this space. Despite their roles and organizational efforts, women are not fully involved in the emergence of modern organizations. For being a source of innovative change, organizational processes must integrate various factors related to the position of women in the community and relationships they undertake with men. The use of a gender perspective in the implementation and consolidation of modern organizations can lead to a sustainable improvement for women and a qualitative shift in their status in the society.

Indeed, policy makers and developers need to know what are these traditional institutions, how they work, who uses them, the place, role and voice of women in the public space, etc. To provide greater participation of women, it would be appropriate to take into account their levels of participation and inclusion in social cohesion, social capital and institutional capital.

Life in the rural areas of the Middle Atlas is often incarnated in its traditional institutions: should we not understand, analyze, study according to gender and integrate them into different development processes?

Behind these questions, a deeper problem emerges beyond the scope of this article: Is the 'classic' form of development projects appropriate to trigger processes of empowerment?

Is the intervention of a development projects as it unfolds always the most appropriate response to the problems and the needs of women involved? Is it still relevant in light of the social dynamics and the changing context?

Such questions can better reflect the complexity of development processes and the systemic environment in which they operate. This dynamic toward more empowerment should be developed based on a multitude of influencing factors identified mainly from culture and respect for the social dynamic that exists in and manages the life of local populations.

The article presents itself as a plea for taking into account sociocultural realities of the areas benefiting from intervention projects.

It is also a call to renew pproaches and methods to take into account gender and the social and cultural dimension of development. And a general orientation to a development more appropriate to people's needs and less destructive of the fundamental principles of cultures and beliefs that give meaning and richness to life.

Notes

1 École Nationale d'Agriculture (ENA), Meknes.
2 Independent consultant, Peru.
3 International Center for Agricultural Research in the Dry Areas (ICARDA), Amman.
4 A.Guédira, Présentation du secteur agricole, Direction de la production végétale, MADRPM, Rabat, septembre 1998.
5 Ibid.

6 Plan Maroc vert: www.agriculture.gov.ma
7 Zahid, F. (2008) Genre et gestion des ressources naturelles 5° Congrès des Recherches Féministes dans la Francophonie plurielle. Rabat du 21 au 25 octobre.
8 L. RABACH: Projet Khénifra: Assister les organisations paysannes sans les instrumentaliser. www.tanmia.ma/article.php3?id_article=9031, date of consultation: 14 june 2011.
9 Y. Abdelaziz (2006) Les lois coutumières et les institutions traditionnels chez les montagnards: cas de l'Anti-Atlas.
10 Is the plural of Azerf (the name of customary law).
11 The foqahas (plural of Fkih) are religious jurisconsults, their primary task is to keep the divine values and Islamic ideals and laws in a society (Chariaa).
12 SAGA is the Social and Gender Analysis of project, the umbrella project of this case study supported by the International Development Research Center (IDRC) of Canada.
13 Sbai (2009) and Maâni (2010).
14 Navez-Bouchanine. F. (1989) Enquête, mode d'emploi. Editions de l'Avenir, Casablanca.
15 Ibid.
16 AGAT: Association de Gestion et d'Aménagement du Territoire.
17 Members of Jmaâ and a few new leaders who emerged from the social life of the douar.
18 Moha, president of AGAT, 2011.
19 We used a sociogram to analyse the relationships between members of different formal and informal organizations. We developed matrices that contain the names of members and all links (for discussion, consultation or other links...) between them. The aim is to identify the most important members and groups and clans who have the power, to identify the density of links, weak areas of interaction networks, isolated or peripheral members.
20 L. RABACH: Projet Khénifra: Assister les organisations paysannes sans les instrumentaliser. www.tanmia.ma/article.php3?id_article=9031, date of consultation: 14 juin 2011.
21 Rapports de mission pour l 'assistance technique a l'unité de gestion du projet, 'Developpement rural participatif dans le moyen atlas central' khenifra. maroc. MAR/B7–4100/IB/2000/2069..

References

Akesbi, N. (2006) 'Évolution et perspectives de l'agriculture marocaine'. Online at: www.rdh50.ma/fr/pdf/contributions/GT3–3.pdf

Dupré, G. (1991) Savoirs paysans et développement, Paris, Karthala Orstom.

Lévi-Strauss C. (1962) *La pensée sauvage*, Presses Pocket, Agora.

Maâni, K. (2010) 'Savoirs locaux, genre et ressources forestières au Moyen Atlas (cas de douar Maâmar)', *Mémoire de Troisième Cycle en Agronomie, Ingénierie de Développement*, ENA, Meknès.

Sbai, M. (2009) 'Contribution des PAM dans les revenus des ménages au Moyen Atlas (cas de douar Maâmar)', *Mémoire de Troisième Cycle en Agronomie, Ingénierie de Développement*, ENA, Meknès.

Tidaoui, N. (2011) 'Genre et organisations professionnelles du figuier à Taounate', *Mémoire de Troisième Cycle en Agronomie, Ingénierie de Développement*, ENA, Meknès.

12 The ripple effect of the SAGA research process

The case of Boudinar community in Morocco

Soumaya Ibrahim Huber[1]

Introduction

> When a pebble is thrown in the water it has a very visible impact – or splash – and then the ripples spread out, getting weaker and less defined as they lose momentum. In the same way, a good quality participatory grassroots process can have a strong local impact...
>
> (Newman and Beardon 2011: 11)

The impact and influence of development research is an agenda that has attained momentum over the last few years. This agenda came together because of two divergent concerns. The first, from the funders of research, draws on results-based management and is concerned with getting value-for-money from research spending. The second comes from the development studies research community, and is concerned with whether research undertaken is 'making a difference' (Sumner 2009).

Researchers might be attracted to the impact/influence of their research because of two interlocking reasons related to notions of accountability. First, accountability to often poor and marginalized people, partners, and communities seeking good change rather than engaging in extractive research that only benefits the researcher's career. Second, accountability to the funders of research matters. This is usually defined by funders as the relevance/utility of the research to their overall objectives, that of the donors as well as the public resources drawn from taxation.

The growing scarcity of resources for research places increasing demands on researchers to provide evidence that the costs of participatory methods are justified by the results. Published studies include little evidence on the impacts of participatory research compared to a non-participatory, counterfactual case (Lilja *et al.* 1998).

This paper offers a review of reflections on the participatory gender research that was carried out by SAGA in Boudinar, Morocco, in the sector of water management. It aims at pointing to the importance of research as an initiative that sets a process going and hence needs to become one of the priority areas of future research initiatives. This review of reflections draws upon research that was carried out based on a framework generated for this purpose in addition to the review of literature and project documents.

The objective of this paper is thus to assess the effectiveness of the SAGA research in conveying gender concepts and to capture the process outcomes from the individual and community perspectives. It aims at measuring the net change in empowerment occurring in the community as a result of the gender research. With measure of change, the meaning of empowerment as coined by Kebeer is adopted and which defines empowerment as, 'the direction of change rather than accurate measurement' (Kabeer 2001: 17–57).

Brief literature overview on impact/influence of participatory research

The aim of participatory approaches at community level is to develop people's own analysis of their living conditions, concerns, underpinning persistent inequality (poverty) to create personal and social change (Newman and Beardon 2011). Expertise in using participatory methods and tools in a gender responsive way has increased.

On the other hand however, there is to a limited extent a growing body of literature exploring the quality, effectiveness and scope of such participatory approaches. To assess impacts, researchers are found to employ a range of methods and disciplines as well as a number of techniques. Positive project impacts often reach beyond economic gains, encompassing human, social, physical, and natural livelihood assets.

Existing literature indicates that impact and influence evaluations are determined by their motives. There are two approaches for impact and influence assessments: the forward-tracking (or supply-side) moving from research to its consequences and the backward-tracking (demand-side) moving from research users to identify potential impact/influences. There are substantial challenges in this area as noted by Hovland (2007: 2) and which probably can explain the relative absence of studies evaluating research impact/influence on end users (who may be the poor and marginalized). The critiques raise concerns that participatory techniques are applied as technical projects rather than empowering political processes and that analysis is limited to the micro level, thus focusing on the quality of participatory development as facilitated at local level.

The same situation holds true for the use of gender analysis: many objectives and expected impacts are attributed to gender differentiation. At the same time, gender analysis is applied with little systematic discrimination (Lilja *et al.* 1998). According to Flintan (2003) the problem here is that gender might have been targeted but without understanding 'gender', which in turn can have adverse impacts, such as: increased gender inequalities and misunderstanding and mistrust between authorities, development organizations and communities, particularly women. Moreover, the brief review of the literature revealed that the literature available examines the research impact on the introduction of a crop, or of transfer of modern agricultural technology, with a gender perspective. A big gap, however, has been detected for research that examines the research impact on the gender dimension in itself occurring as a result of gender research, such as understandings of gender discriminatory actions and perceptions.

The larger development challenges of impact/influence research

There is an assumption that research expenditure has, in the long term, a higher value added than alternative uses of development funds such as aid funding. Many studies, as listed by Surr *et al.* (2002: 8–9), purport to demonstrate the strength of research in reducing poverty and note that the cost of lifting one person out of poverty through agriculture research was US$180–190 per person, compared with US$2,304 per person for lifting one person out of poverty through aid spending in general.

The main research hypothesis of this investigation is that SAGA research in Boudinar on the use of water management which adopted participatory gender sensitive research tools in its data collection has affected gender sensitive understandings among the participants of the Boudinar community.

The basic question we ask is: What are the ripple effects of the participatory research on the gender dimensions among the Boudinar community? The research question is important as it helps us to understand how the research process benefited the different stakeholders in Boudinar. It helps us to address both the equity concerns, by focusing on the benefits accrued to different beneficiary groups (male and female members) and the effects of the participatory process on human capital.

The development implications of the research question have two aspects. Firstly, it addresses research practices in mainstreaming gender in environmental work through the emphasis of empowerment of women in decision-making and gender balance in meetings. Secondly, it strengthens the bottom-up approach and is an affirmation of the centrality of gender equality to human rights.

Study area and methodology

This research was carried out in the Boudinar area located in the northern part of Morocco, and west of el Nadur territory. The Boudinar community is one of the oldest ones in the region and is considered as the mother community for the other four communities which together compose the Timsaman tribe; namely Timsaman, Trocot, Awlad Imghar and Beni Marghaneen. The locations visited were Boudinar center, Bouma'ad; the school of Maguia and Ishta school as well as Hamout house.

The research adopted the qualitative approach. The information collected focused on three main categories, namely: 1) the pebble (the research design and process with tools), 2) the actor throwing the pebble and the mode of throwing (the researchers) and 3) the ripples resulting (the result, output and impact), or the instant and the latent ripples.

Five tools for data collection were selected:

- **Individual interviews** Semi structured interviews with community members, (seven interviewees: one male and six females) were chosen because one-on-one interviews can successfully facilitate participant–researcher knowledge exchange, catalyzing further knowledge sharing with community members.
- **Key informant interviews** (two males and one female interviewee) were selected because the key informant is usually found to be working and living in the community yet originally from outside the community. Thus key informants are found to have more of an insightful knowledge yet remain neutral due to their outsider nature.
- **Group interviews** (two focus groups with female interviewees) This tool allowed for participant–researcher knowledge exchange as well as participant–participant knowledge sharing. The tool was successful as it allowed for a collective discussion of problems, knowledge consolidation, opinion sharing and realization of existing capacities.
- **Observation** as a tool was selected because it provided verification for what the interviewees conveyed in their talks, such as their level of engagement and interaction with other group members, their interaction towards the other sex in terms of respect, submissiveness etc.
- **Review of literature and secondary material** was essential to complete the background picture. Project documents as well as literature about the relevant experiences undertaken in similar initiatives were reviewed, all of which helped in the clarification of the research questions.

The field visit to the Boudinar community lasted for one day and stretched from 10:00 am til 8:00 pm. The researchers who were part of the team and belonged to the first two groups, as will be presented later, were met individually on consecutive days after the field visit.

Sample and its selection

The sample in this study consists of 19 interviewees and is composed of 16 females and three males. All male interviewees are in the age category between 31 and 55 years and two of them are employed while one works as a farmer. Among the three male interviewees there was only illiterate.

Half the female interviewees are in the age category of 16 to 30, while the other half are in the age category of 31 to 55 years. All female interviewees with the exception of three are unemployed and illiterate. All participants in the sample have been selected according to availability during the ad hoc field visit. The only criteria made for the selection necessitated that the interviewee had participated in one of the previous SAGA initiatives.

All meetings were gender segregated to suit local cultural situations. Group activities used existing community groupings to limit intrusiveness and difficulties with logistics. Groups generally included four to eight individuals.

Individual interviews lasted for one hour or slightly more due to the translation from Amazigh language into Arabic. The group meetings lasted for two hours. All meetings were arranged in the same location for the community members. Group meetings were conducted in schools and individual interviews were conducted in private homes.

The information for the impact assessment was collected based on a framework that was designed for the purpose and is portrayed in this section. It includes the methodological approach and the accompanying tools.

The 'pebble' refers to the input side and addresses the question of, how does the project's scope and approach influence the impact? Here the focus is on the research design and the delivery process of the research such as research content, tools applied. It also attends to the researchers as the change agents and their roles in the process. As they are the one to drop the pebble, how did they drop the pebble? What arrangements were made in terms of space and time? Where were the meetings carried out and when and for how long?

Table 12.1 Summary of methodological approaches/tools framework

What is the pebble and how was it thrown? INPUT How was process designed and information obtained?	The Ripple effect **First ripple** OUTPUT How was information captured?	**Latent Ripples** IMPACT *How was knowledge used?* Indicators On individual level	On community level
The role of the facilitator/researcher or change agent	Which method was preferred and why? Which situation, setting was the best and why?	Awareness about gender roles Usefulness of tools	Knowledge sharing Conflict resolution Living conditions
Space and time and arrangement of participation	Women's/men's participation in the research	New participation opportunities Voices heard	Networking with locals/neighbors
Identification of local definitions and existing GENDER gap	Whose voices have been heard? Whose knowledge and opinion counts? What was the community contribution?	Self-image	Community members' notions and practices of gender roles Participation of women and men in community affairs Level and quality of participation
Tools: secondary resources and interviews with primary change agent	*Tools*: interviews with primary change agent and focus groups	*Tools*: Interviews with individuals and focus groups	*Tools*: interviews with male and female interviewees and focus groups

Identification of local definitions and existing gender gap: how does the community define gender roles, inequality and poverty? What kinds of strategies did the researcher adopt to overcome or shift the power imbalances occurring during the research?

Who threw the pebble? Who are the stakeholders in the impact assessment and what are their impact interests? The change agents comprised several persons who played different roles and functions which can be grouped into four change agents according to the role and responsibility they played. The primary change agents 1) included the head of the project, or the team leader and two female researchers, who acted as researcher and facilitator for the workshops and conducted interviews. They were the ones who were basically responsible for the prime decisions though of course to varying degrees.

The second change agent group 2) was composed of two females both of whom had gone once or twice to the community to perform a one-shot assignment. The author, appointed by ICARDA to coach the initiative, is one of them. She conducted one day training in gender analysis for a small group of female community members (16) which was also attended by members from a women's NGO from the neighboring community. The consultant also conducted case studies with female community members.

The third change agents 3) were composed of key informants of the Boudinar community, in the sense that they are not necessarily from among the community members, but live there because of their work. This team comprised two males, the head of the community and his vice president, and two female employees. One of them works as the community secretary and as researcher who had accompanied the author during the case study interviews as well as the impact assessment research. The other female candidate in this group is from the community and teaches adult education for women in Boudinar.

The fourth agent 4) comprised the male and female Boudinar community members who were involved in the research project as they were invited to the workshops or interviewed.

So what is the pebble and how was it thrown? The very first intervention for SAGA started in August 2009 with a workshop that was conducted by two of the female primary change agent group. Later that year, after the initial workshop, the prime researcher who was also conducting the field work for her MA started to travel around the five douars of Boudinar to allocate the water sources and their suitability and usages. She was accompanied by one male community member and another female community member, who works as a literacy trainer in Boudinar.

The workshop series then continued throughout 2010 and were conducted in the five douars that had been selected by the primary agent group. In four of the douars and the commune, the meetings were conducted in the local schools and in one douar only in the mosque, as there was no school in the area. The meetings were conducted separately for men and women. While for men it was decided to have the meetings during the day, for women the meetings were conducted in the afternoon, as this was perceived to be more accommodating for women's time schedule. Thus the meetings were held in a row and successively on the same day. As soon as the men's meeting ended, women were ushered to come in.

During the same year, 2010, individual interviews with questionnaires were conducted by the prime researcher with approximately 20 female and male community members. The researcher also lived for one week in the Boudinar community, spending each of the nights in that week with a different family in the community. Towards the second half of the year, this primary researcher role ended.

In December 2010 the one-day gender training course was conducted by the SAGA newly hired consultant coaching the initiative. The consultant conducted as well as four case studies with female community members a pilot, testing for a larger case study initiative that was undertaken later on by a team of local researchers. The case study research continued until March 2011.

Identification of local definitions and existing gender gap. Prior to any workshop or work performed in any community, the prime researcher met with the eldest male member in that douar to obtain from him the necessary information about his community members aiming of course to learn about the gender gap and the power imbalances such as: the difference in accessibility, distribution of resources, access to services and various roles and activities performed in the community. The identification of the interviewees for the individual interviews was left entirely to the male head of each douar, who selected the members he found applicable to the requests of the researcher. In a few cases he ushered the researcher to several candidates and left her to do the final selection of the interviewees. The invitations for the workshops were announced via the microphone of the mosque for the information to spread as far as possible.

The ripple effects: what are the most important impacts to be measured? This refers to the output and focuses on how the information was captured and influenced the community. Here two types of ripples are identified. The first ripples occurring immediately already during the research and include aspects such as women's/men's participation in the research, whose voices have been heard? Whose knowledge and opinion have counted? What was the community contribution? Which method was preferred and why? Which situation or setting was the best and why? The second or latent ripples point to the outcome and how the information was used by the community or affected them later on. In order to assess these, two different types of indicators were chosen: those that assess personal changes at the individual level, and those that look at community level changes.

The personal indicators were to assess awareness about gender roles – their notions and practices of gender roles, participation opportunities, voicing their interest and self-image, self-esteem[2] and usefulness of meetings.

Group/community indicators were used to assess: levels of knowledge sharing through the meeting or afterwards with family members or neighbors, conflict resolution in water management as a result of meeting, living conditions and how participants felt about it after their meeting, networking with neighbors as a result of meetings and integration in community and community members' notions about men's and women's participation in community affairs.

How do we judge the success of the project? After the two rounds of field visits, a content analysis was performed and the information was categorized according to the designed framework and sorted according to the set of indicators as shown in Tables 12.1 and 12.2.

To analyze the level of participation of the actors, Table 12.2 was designed. It portrays the entire research undertaken by the SAGA research project and groups that into nine activities. The activities included: planning, training, design of research, data collection, data analysis, data presentation, report writing, use of evaluation findings, and assessment of impact evaluation. In this table the first change

Table 12.2 Level of participation of the different actors/groups in the research process

Stages	*Team leader*	*1st change agent*	*2nd change agents*	*3rd change agents*	*4th change agents Community/ beneficiaries*
Planning	Objective planned, based on previous projects implemented there H	–	–		–
Training			H	M participated and organized	L few and only female participants attended
Design	H	L		M Select logistics, criteria	
Data collection		H	M	M participated in interviews	L were interviewed
Data analysis	H	H	M	–	–
Data presentation	H	H		–	–
Report writing	H	H	H	–	–
Use of evaluation findings	H	M	H	–	–
Assessment of impact evaluation	L	M	H	M participated and conducted interviews	L participated in focus group and interviews (reflections)
Total	6	6	6	4	3

Key: H = High, M = Medium, L = low, – = 0.

Source: Adapted from Roy Abbes (1998).

agent is divided into two columns for clarification purposes only, as the team leader was found to have played a special role within this group. The participation level of each change agent group was distributed along the rows and columns. The level of involvement is measured quantitatively through the frequency or amount of intervention as well as qualitatively through the degree of involvement, meaning high, medium and low. For example; high involvement means having been always involved and in control of the activity.

The review of the participatory interventions of the research project reveals that there were slight attempts to be participatory and involve the community members. In terms of frequency, the first two change agents (including the team leader) have had six interventions in the project, while the third group had only four and the group of beneficiaries only three.

The intensity, or high level involvement, is found only among the two first change agents and disappears completely from the last two groups.

Towards the end of the columns (groups 3 and 4) the level of participation is not only weak in frequency but also is of a medium type. The data analysis occurred through the following means:

Data validation

After collating the data, interviews were held with key informants to discuss the findings in order to verify them as well as to fill the gaps resulting. The results of the discussions have shown that the key informants found the assessment itself a positive learning process.

Triangulation

To complete the verification process, the information collected was clustered and labeled in the pertaining issue once it was confirmed from three different sources, such as individual interview, focus group, key informant, the authors' (my) observation, change agents and or review of secondary material. A good opportunity arose when interviews with key informants could be done at the end of the research process which allowed further discussion with local communities.

The two control approaches

Judging the success of the research process through two control approaches, namely

Reflexive

Participants who received the intervention were invited to compare themselves before and after receiving the intervention. They were also requested to cite examples (indicators) in order to prove their point.

Generic

Here the impact of the intervention of beneficiaries is compared with established norms about typical changes occurring in the target population. This was accomplished through the literature review (Rossi and Freeman 1993).

Finally, when researching the complex dynamics of influence the question arises of how to assess the contribution of research to a certain attribution or outcome? This research sought its findings through the investigation of each tool adopted and its influence on the community members, thereby adopting a more specific approach in the investigation in order to arrive at a specific outcome. Hence it strongly maintains that tracking research impact and influence is usefully thought of as demonstrating contribution rather than attribution (Sumner, 2009).

Results and discussion

What are the ripples of the participatory gender-sensitive research (cause) on the gender understanding among the Boudinar community (effect)?

The instant ripples

Change agents are very influential in introducing a new gender role model in the community

The primary change agents, (the female researchers) played a vital role in introducing a new role model for women, because of their personalities and their experience and achievements. Being: 1) a young woman, 2) educated and 3) emancipated because they travel on their own, and 4) decent and respectable, prompted the female community members to perceive them as their role model, to whom they aspire and want to be alike. In a similar vein, the female community member responsible for working in the Boudinar commune was also highly recognized and appreciated by the community members, because she had attained an advanced educational degree and mastered both the local Amazigh language and wears their traditional costume.

Participatory tools adopted began to close the gender and age gaps

Despite the fact that there was a gender variation in the tools preferred, as male community members liked using the village maps where they had to allocate the natural water resources and women preferred the daily routine diagram to share their duties and responsibilities, the gender gap began to be closed. This is because through that exercise, subtle attempts were made, and both women and men in their segregated meetings, were invited to talk about the activities of each other based on their observations. The discussions in the men's workshops, however, touched only slightly on women's roles, as some men consider it inappropriate to talk about their female family members in public. Other reasons why men did not talk fairly about women's roles were that men consider it very normal that women do all the tiring jobs, because it is their duty to do so, and because men have to work even harder to gain the families income.

Asking men to describe a woman's day has in fact increased men's understanding of women's work. As maintained by numerous responses of local key informants, this discussion could not have happened by itself, primarily not among men, as it is considered shameful for a man to talk in public about the women of his household and their performances.

The age gap began to close as well as elder men and women showed similar interests to share their knowledge about the historical aspects of the community and its resources with the younger generation.

Meetings opened new opportunities for participation for women, yet adhered to the conventional gender code of ethics

In general, according to a male key informant, women revealed a much higher awareness because they were eager to attend, which proves that these workshops were grasped by women as opportunities for them.

There is consensus among the respondents that there was a substantial women's participation throughout the workshops, though men's participation can be estimated to have exceeded women's participation and was in a ratio of 60% to 40%. This ratio is a fairly good one, given the fact that in Boudinar the prime reason for women's absence is because women are either busy or uninterested to attend. Factors affecting women's participation, however, are found to be due to social status, age and education. Accordingly young married women as opposed to elder married women were not allowed by their husbands to join the meetings. Only unmarried women of any age were usually participating in the workshops. Thus the overwhelming female participation was found to be unmarried women, but predominantly from young and educated girls.

In Boudinar in general it is considered inappropriate to conduct gatherings with a mixed audience. It is only on rare occasions where men and women convene, but usually in these rare cases the audience has been an educated one. The female researcher, however, could attend the men's meeting and in fact participate in the facilitation. Likewise in one area only, in Tizza, the male chairperson of a local NGO used to attend the women's workshop occasionally. He was allowed to do so, because he is from the community and his NGO had worked extensively with the local community before, so they had associated him with serving the community and hence confided in him. No male facilitator otherwise was allowed to do so.

While the various approaches were addressing both men and women, the gender training remained confined to women as there seemed to have been a misunderstanding among those responsible in the community that only female community members were targeted, and hence only women were invited.

Ultimate ripples

Female participants felt empowered with a heightened self-image and enhanced awareness

According to the researcher, at the end of the workshops, participants were always cheerful and self content. They felt very proud as if they were attending school, especially as the workshops were organized in the village school. Several indices attest to their increased power and self-confidence.

- Several women as a result of the workshops got encouraged and requested from the facilitator that she arrange for them a literacy training class. In other cases, participants asked the facilitator who worked as an adult literacy teacher to include them in her literacy classes.
- One woman went after the workshop to meet the head of Boudinar and requested him to provide projects for women and specific educational classes for women.

Women's own words reveal the increased broadening of options:

> 'I got the power due to the people who came from outside and I felt that I have been released from prison.'
>
> 'We came out of prison after we have been marginalized and deserted like the animals in the mountains.'
>
> 'We used to watch people discuss on TV but now we also do like them and we meet and discuss.'
>
> 'We were always happy when people were coming from outside to see us.'
>
> 'I got very happy and felt that there is someone who cares about me.'
>
> 'I want to be educated so that I can understand everything that happens in the workshop.'(Fat)
>
> 'It gave us the opportunity to express ourselves so that our ideas can be delivered to the policy makers.'

In the women's own words increased self-awareness is revealed:

> 'I learned to think in a new manner and when they were doing the exercise I realized how important it is to be educated.' (Maa)

Meetings exposed existing discourses which involved inequalities such as gender and class relations. The SAGA research project was able to help create spaces and opportunities for less powerful women to negotiate with other groups and to exercise choice. Throughout all the workshops, there was a very dynamic interaction, though probably usually towards the end of the workshop, as stated by the female researcher. She recalls how silent the workshops started and then how vivid they became towards the end. All participants were always encouraged by the facilitators to share in the discussion with their opinions. Facilitators also ensured that the less active ones join the discussion by asking them whether they approve of what is said. Every workshop, according to the facilitator, the audience questioned her about the output of their discussion and where it will lead to.

> 'I now have understood that I perform two roles and that I have a double load.'

A sense of empowerment but then what?

This heightened sense of awareness and empowerment which female participants conveyed and the inspiration they always felt when going to the meetings, began to vanish as some women felt confronted by their own reality afterwards.

> 'We spoke about the role of women and that of men and that there are activities which we cannot do but we do not know the solution for this.' (M)

> 'I feel deplorable after the meeting because there is nothing I can do and there is no one to help.' (F)

> 'We got to know that women do not have time to think about themselves and we wish that they stay beside us so that we can do something which we did not do before, such as income generation, instead of only working with no profit for others.' (R)

The same feeling of support was also conveyed by a male member who said: 'There is someone who thinks about me and someone who bothers about me.' (Yo. Moh)

Ripples at the community level

The meetings initiated platforms for knowledge sharing among male and female community members

During the workshops issues about climate change and water management were presented and discussed, which encouraged many female and male participants to implement the information they received.

> 'It explained to me the impact of climate change on my own life. I as a result started thinking how I could deal with it in my life practically and I talked with my family members and neighbors about it.' 'We agreed in that workshop that we will plant trees and it is due to the workshop this decision.' (Male community member)

Role of women in conflict resolution started to be publicly addressed

During the workshop the location and usage of various water sources were discussed and inevitably of course the disputes around them came up during the talk. The opportunity itself to discuss this was appreciated as expressed by one of the female participants:

> 'We are very happy because we spoke collectively about our problems and because someone came to us.'

Some of the discussions were successful as well in ameliorating between the different disputed partners as expressed here:

> 'Our elder generations had problems or conflicts around the land, and we were not directly involved. Thus during those meetings it was a great opportunity for these concerned parties to meet and to get together again.' Some of the workshops, however, created disputes among the female participants during the workshop about a water source, for example, that had been seized by some families, and hence led to a water shortage. Often these disputes were not resolved at the end of the workshop and were carried on later among the community members.

> 'No on the contrary sometimes rivalry was created after some participants had expressed their opinion and the latter was not welcomed by everyone. Fierce negotiations started about the rights of usage or old conflicts.'

Workshops provided male and female networks for local neighbors

Participatory workshops provided many opportunities for local neighbors to meet and hence male and female networks were either created or simply revived.

> 'I feel happy because we got to know new people.'

> 'We used to meet together and instead that each of us carries her own concerns alone, we all carry it now together as one group.'

> 'We got to know each other more closely.'

Also for men the workshop meetings provided an occasion for getting together; 'Those meetings were an occasion for us to meet.'

The workshops contributed to various changes in the community's perception of gender roles, which varied between acceptance, recognition and threats

In Boudinar, there is an overall increasing acceptance among men for the role of women.

> 'Over the last few years women in Boudinar have become emancipated and they no longer stand hand bound. It is now accepted that women have their

own projects, continue their education, are involved in associations and even head women's associations.' (Male key informant)

According to the respondents' feedback, the change incurred through the research process varies between: 1) increased acceptance of gender roles, 2) recognition of women's roles as well, 3) feelings of threat on the part of women and men alike.

Increased acceptance of gender roles

Both men and women revealed acceptance of the complementarities of the gender roles. According to a male interviewee:

> 'Women have become more interested and are always eager to know what happened in men's workshops.'

Two other women expressed this perception in their feedback by saying:

> 'Even though all the meetings were conducted separately, there is now increasing acceptance for mixed meetings and also acceptance for men who come from outside the community.' (F.S)

> 'If men would not have talked, women would not have been able to do something on their own. Everything has to be done equally between men and women and if something has to change they both have to work together.' (Hd)

Increased recognition of women's roles

Female members expressed their perceptions about men's or community's increased recognition of women's roles as follows:

> 'Men have opened up and they are accepting us more.' (Fa)

> 'Now the community comes and asks us about our needs and we also agreed that water is a priority and hence has to be repaired.' (MYM)

Men are seen to confide even more in women. According to (La) after one of the meetings one man from the community came to her and told her:

> 'We heard that in your workshop you expressed yourself the most, and therefore I delegate to you the matter and you do whatever you find appropriate. You can either repair the water path or what you see as the best.'

Feelings of threat

Still for other women, men started to recognize women because they fear them,

> 'Men began to be more concerned about women because women now threaten them in their roles.'

Men are believed to recognize women's role when they can use women's voices for their own benefit.

> 'When the spring was repaired, men were the actual ones to gain but for us everything remained the same we only supported men when we expressed our needs and thus they obtained what they want.' As summarized by the female key informant:

> 'Even though the workshops as well as the training organized contributed to changes in the community's perception, the change is by far not widely noted. A more inclusive awareness raising is needed for all community members, namely with youth and many more women. We are still living in a male society and hence much more support is still needed to keep the momentum and avoid threatening backlashes.'

Within the women's workshops, an interesting scheme became obvious whereby women decided informally among themselves on who should be the spokesperson to represent their opinion in public. During the field visit in December of the author, the same scheme was observed. There were two women that the other women used to whisper their opinions to, so that they could share it in public.

> 'I can only say if women are hesitant to talk then they are poor.' (Ma)

Men's workshops on the other hand revealed much power games, and while women were making suggestions that had their family's welfare in mind, men were found to be seeking their own interests.

More marginalized community members, however, were not included, as revealed by Fatima, a handicapped woman (42 years) who fell from the donkey when she was 35 and now walks with crutches. According to her,

> 'I was not informed because people do not tell me. Because of my handicap, people think that I will not be able to attend.'

Discussion on how these results explain and expand our understanding of impact and influence of research

The results obtained expand our understanding of impact assessments either as confirmations of information already known or as results creating and widening understanding. In this section we will make use of the two research processes: the initial one conducted by SAGA as well as the research for the impact assessment, in order to discuss if there are any ingredients for impact/influence or factors that support or impede the impact/influence of development research. Some literature has tackled this and mentioned a multitude of complex factors likely to influence the impact or influence of development research. These include those factors that the researcher has a degree of control over such as methodologies, engagement with users, to those outside her or his control, but attention to which can lead to opportunities, such as taking strategic advantage of the political and economic context during the period of the research. There is no single recipe for impact or influence but there are ingredients that stand out as being important. Notwithstanding the diversity in the literature, three domains or clusters of factors emerge which focus on the underlying notion of the social life of an idea – the factors are divided into these domains: 1) content and processes of knowledge generation and translation; 2) connecting with and creating champions of impact (Cornwall 2003), influence, 'knit-working' and change; and (3) context, 'strategic opportunism' and luck.

According to the findings of the research assessment, we will here discuss those factors and relate how they interrelate with the results to help expand our understanding.

Content and processes of knowledge generation and translation

This domain refers to the engagement and participation of users of research at the outset and during the research. The findings reveal that the content and process can be supported if the research embraces gender and diversity issues from the start by having a gender-sensitive research plan with gender-sensitive tools and approaches. For example in the case of the SAGA research process, the tool of living in the community was found to be a very practical approach to assess their real problems and needs. Factors limiting the impact of knowledge creation were found to be the absence of a baseline study or gender analysis framework according to which a gender-sensitive criteria for selection of sample could have been done. Findings of the research assessment reveal there was gender blindness in the selection of female respondents. Women have mainly been grouped according to their geographical origin (commune) rather than to any other factor differentiating them (such as age, class, occupation). In future program activities, particular importance should be attached to supporting initiatives ensuring the accountability of women members to all the other women in their community.

In a similar vein, the results obtained through our research methodology confirm as well that the women's mutual support was possible because of the interactive approach rather than an extractive one. Importantly, knowledge generation was a two-way process. Inviting participants to tell stories in an attempt to verify the indicators created a relaxed and highly personal context and also provided an opportunity for participants to ask questions of the researcher. In this way, the assessment became once more an important platform for raising awareness of gender issues in the community. Furthermore, through the course of discussion, links between water management-related problems and more general problems (such as loss of traditional knowledge) were clarified for participants as well as the researcher. In this way, the research was interactive, not extractive.

Champions of impact, influence or change agent

The findings as discussed confirm what is already known, namely that the impact /influence of the research process can be increased with a well-structured research design and gender-sensitized researchers. As important or even of utmost importance in any research or pebble thrown is also the manner with which the pebble is thrown. Cornwall, for example, already pointed out that the core of the 'problem' is not that of participatory methods per se, but rather the use of the methods and the assumptions of project workers. Our results reveal that the role of the researcher is not only important, because of how they raise the gender issues or what they think gender means, but also because they do in fact introduce a new gender role and can become a role model for women to look out for. Thus the values of individual facilitators and their presence will influence participatory processes and the extent to which women's rights are supported (IIed 2011).

Context, 'strategic opportunism' and luck

With relevance to the factor of context or what Gladwell (2000) has coined as 'power of context' because influence and change to results often need a conducive environment. Thus research must not work in a vacuum but understand the relationships and linkages between the projects and 'external' factors including social, political, cultural and economic pressures and/or change. If research is to have impact, then adaptability, flexibility and a long-term focus are vital. Accordingly issues such as gender equity cannot be addressed overnight but require commitment, time, resources and sensitive, well-informed interventions or support (Flintan 2003). Hence linkages must be made between the local and the national level as well as the international level to introduce the community to what is going on elsewhere, from which they can possibly learn. For example one such lost opportunity could have been to link the gender-sensitive research to the national Initiative for Human Development which has among its objectives the improvement of women's livelihoods and their social status. This initiative lies within the realm of the national strategies implemented in Morocco since 2005. In Boudinar some villages had received goat breeding projects run by women.

Another external factor impeding the influence of the research project, but that went unnoticed, was the severe economic circumstances that had increased the expectations of the community members for a material project to result from the meetings they were invited to. These severe conditions made people want to see the influence of the success on their daily life. According to one of the female respondents (Hd):

> 'Nobody follows up, they only come and ask questions and collect information but they never return again to inform us how far they have reached with the information.'

In this case a demonstration project in the village might have helped ameliorate this concern.

One last aspect that also had a limiting effect on the research impact, was the high illiteracy rate especially among female community members and the elder ones in particular. In future work, gaps such as these need to be studied before the research design is crafted so that the appropriate tools are selected accordingly which match the circumstances and increase the project's effectiveness.

Discussion on how the research findings compare with other findings in the literature

The findings of the impact assessment are very much in alignment with other findings in the literature. In this section we are going to focus on the most important ones, namely those related to the direct impact on women's empowerment. Our findings revealing that women's groups initiated through programs or research can lead to increased self-image correlate totally with many studies. Findings in the literature state that women find a voice and strength through collective action. Hence promoting women's participation through women's groups proves highly successful. A gender assessment of a WWF-supported project suggests that because of the groups:

> the unity of women has increased and strengthened their own self-image as they now feel that they can achieve what they intend to do on their own. The members felt that they [have] a place to share and express their experiences and difficulties…now they are confident to talk with others freely and are able to voice out their opinions.
>
> (Samanata 2001b: 5 in Flintan 2003)

Other findings reveal that research programs have enabled women to discuss an issue – as in our case water management – that was previously considered within men's exclusive domain. In so doing, women have demonstrated the important contribution that they can make through their skills and understanding of the issues involved. They also indicate that activities have increased women's awareness of and confidence in their abilities and knowledge. For instance: women's self-confidence has been increased by overcoming their fear of acting outside culturally sanctioned (patriarchal) norms. This confidence or awareness effects the first step to women's empowerment towards their own needs and priorities. Moreover, research programs promote women's sense of belonging to a distinct social group with shared interests, facilitating individual and collective action.

More generally, the activities may have wider gender implications in other areas of social and family life. Activities may contribute to broader social and cultural change and have positive spillovers on other areas of gender concern (Hamilton and Dama 2003).

Implications for development policy and recommendations

In this section the recommendations with relevance to development and policy implications which are based on the research findings are described.

As demonstrated, gender-sensitive participatory research can be both a goal and method of change, if it first refers to the society without the monopoly of political, economic and social powers in the hands of the elite; and second, if participation is used as a means that serves to develop the capacity of those presently excluded in order to identify and express their needs and contribute to solving them (Galina [n.d.]: 6).

Research conducted will have a lower impact on the gender dimension of the communities researched if it fails to approach gender issues in any strategic way. It is then a lost opportunity. Accordingly: research staff need to have, prior to their intervention, received gender training and a gender analysis checklist to take a more comprehensive, informed and successful approach to overcoming problems and inequities (Flintan 2003). They need a plan for participation to ensure the integration of all stakeholders at the start of the research process so that they are involved throughout all the stages from the design etc. Defining who participates through conducting types of gender analysis is potentially a determinant of the impact of the participatory process on the research results.

Prior to the start of any research process, it is very crucial to decide on the research context as it is strongly linked with the methodology selected. A 'research only' context as opposed to action research where decision-making stages are the prime focus, influence the tools used as well as the level of participation requested and hence the research outcomes and participants' outcomes. Each type of research necessitates the selection of tools and to match them with the community members. The research context also emphasizes if a research process will be interactive or extractive.

It is important to have a vision of what the project is ultimately to achieve: extractive research, dialogue, interactive or something else? (Flintan 2003). This should be realistic and achievable given local contexts and circumstances (including gender issues). The pathway to this vision should include a viable and well-planned exit strategy. This should focus on sustaining activities and maintaining (and indeed expanding) the links created.

Using a participatory approach, community benefits can be maximized via the research process itself and even without material outcomes. Community learning and empowerment can be facilitated through collective discussion of problems, knowledge consolidation, opinion sharing and realization of existing capacities. 'Interviewing' methods emphasizing participant–researcher knowledge exchange or larger group activities aimed at collective participant– participant knowledge sharing need to be tested or combined in order to achieve this goal (Warrick [n.d.]).

The way forward

In general, the indicators of impact/influence are under-researched and it is clear that there is a gap in qualitative (as well as quantitative) indicators that adequately account for the impacts and potential impacts as well as more deeply embedded indicators for assessing less visible change, for instance in attitudes and personal development (Flintan 2003). Although researchers using participatory methods have observed the success of the approach in a variety of fields, and documented results in a number of case studies, there remain insufficient approaches to participatory research and gender analysis (PRGA). Is PGRA most effective for different types of technologies such as knowledge sharing or empowerment? (Lilja *et al.* 1996).

New emerging questions through the findings

Working to increase the participation of excluded groups, both men and women, is a political rather than technical project. Participatory methods help to effectively use spaces to strengthen the voices and perspectives of disadvantaged groups. However, the critical questions remain, such as the questions of whose participation, in what and for which ends remains critical (Kanji 2004).

In a similar vein what is the social differentiation and representation among women? In the case of SAGA for example, preliminary observations suggest that there was a bias towards educated women members of women associations during the training. Also during SAGA research, results reveal that women have mainly been grouped according to their geographical origin (commune) rather than other factors such as social status, age, ethnicity, or marital status.

Are the results sustainable? It is important, however, to recognize that changes in attitudes and perceptions that are entrenched in local culture (and often internalized by women themselves) require long-term commitment and processes because silent resistance and gender stereotypes may require a longer time to be overcome.

As mentioned before, many studies purport to demonstrate the strength of research in reducing poverty (Surr *et al.* 2002). So if research can reduce poverty can it also reduce gender inequalities? Tracing the impact/influence of research from research outputs to reductions in poverty and inequality is difficult (Sumner *et al.* 2009) as highlighted before, but methodological attempts enabling this would definitely be worth trying.

Participatory research is the start of a change process. Further on, if throwing the pebble is a big act in itself that deserves solid planning and action, then the ripples occurring from that act are even more important. They are the resonance of our research process deeds and therefore deserve due attention and are certainly worthy of recognition and wide dissemination.

Notes

1 Consultant, Cairo.
2 Here we have picked up on the notion of 'self-esteem' which, in psychology, refers to self-love, self-perception, self-confidence, recognition of one's own abilities and others' recognition of one's abilities (others' appreciation).

References

Cornwall, A. (2003) 'Whose Voices? Whose Choices? Reflections on Gender and Participatory Development', *World Development* 31(8): 1325–42.

Flintan, F. (2003) 'Engendering' Eden: Women, Gender and ICDPs: Lessons Learnt and Ways Forward', Summary Document. Wildlife and Development Series No.16, International Institute for Environment and Development, London.

Galina Chus-Moulaye (n.d.) Oooaiu Environmental Governance Student's profile Participatory and Gender Approach Atlantic International University.

Gladwell, M. (2000) *The Tipping Point: How Little Things Can Make a Difference*. Little, Brown and Company, Boston.

Hamilton, L. and Dama, A. (2003) 'Gender and Natural Resource Conflict Management in Nioro du Sahel, Mali'. IIED Issue paper no. 116.

Hovland, I. (2007) *Making a Difference: M&E of Policy Research,* Working Paper 281, London: ODI.

Kabeer, N. (2001) 'Resources, Agency Achievements: Reflections on the Measurement of Women's Empowerment', *Development and Change,* 30: 435–64.

Kanji, N. (2004) 'Reflections on Gender and Participatory Development', Participatory Learning and Action 50, October.

Lilja, N. A., Ashby, J. and Sperling, L. (eds) (1998) 'Assessing the Impact of Participatory Research and Gender Analysis', The Consultative Group on International Agricultural Research (CGIAR) Program on PRGA for Technology Development and Institutional Innovation, 2nd International seminar in Quito, Ecuador 6–9 September 1998.

Meinzen-Dick, R., Adato, M., Haddad, L. and Hazell, P. (2004) *Science and Poverty: An Interdisciplinary Assessment of the Impact of Agricultural Research, Food Policy Report,* Washington: IFPRI.

Molas-Gallart, J. and Tang, P. (2007) *Policy and Practice Impacts of ESRC Funded Research – Case Study of the ESRC Centre for Business Research,* ESRC.

Newman, K. and Beardon, H. (2011) 'How wide are the ripples? From local participation to international organizational learning' *Participatory Learning and Action* 63.

Surr, M., Barnett, A., Duncan, A., Speight, M., Bradley, D., Rew, A. and Toye, J. (2002) 'Research for Poverty Reduction', Department for International Development, Research Policy Paper, London: DFID.

Sumner, A., Ishmael-Perkins, N. and Lindstrom, J. (2009) 'Making Science of Influencing: Assessing the Impact of Development Research', Working Paper 335, Institute of Development Studies at the University of Sussex, Brighton.

13 Capacity building effects of the social and gender analysis project

Approach and lessons learnt

Andrea Pape-Christiansen[1], *Malika Abdelali-Martini*[2] *and Aden Aw-Hassan*[2]

Introduction

The Middle East and North Africa (MENA) region faces enduring and escalating challenges due to poverty and environmental insecurity, exacerbated by social inequalities and inadequate policies. These development challenges require strong multidisciplinary research that integrates social and institutional issues alongside natural science dimensions. However, the region is hindered by an inadequate capacity to integrate social and gender issues in environment and natural resource management, particularly among National Agricultural Research Systems (NARS), limiting the effectiveness of research efforts.

Extensive consultations with researchers from around the Middle East revealed key challenges to the integration of social and gender analysis (SAGA) in ENRM research in MENA. These challenges include: lack of integration of social and natural science in institutions; limited technical capacities in SAGA; the added value of SAGA unrecognized by managers and natural science researchers; limited education and training programs (lack of integration of interdisciplinary approaches and SAGA in education systems: training on gender is often short-term with no field-based support). The challenges continue due to an absence of regional resources on SAGA in ENRM (absence of documented success stories, case studies, training materials, resources on methods, and tools in Arabic or grounded in MENA experience); missing or weak accountability (poor monitoring and evaluation, lack of follow up by donors and lack of incentives or personal/institutional commitment) and gender-blind policies (including a lack of funding for SAGA).

The project took these challenges as a starting point for its design. Drawing on a review of lessons of social assessment and gender work from around the world, IDRC and ICARDA agreed that any approach to strengthening SAGA must simultaneously tackle four main components: 1. strengthening research quality by applying SAGA tools, 2. capacity building for the use of SAGA tools, 3. organizational support, and 4. communicating SAGA research results. Capacity efforts were designed to assist with all 4 aspects.

The project

The goal of the project was to build capacity and generate innovative practices and methods that demonstrate the value-addition of social and gender analysis (SAGA) methods in action-oriented and transformative research. The project had four specific objectives – the component dealing with capacity building is highlighted below:

1 Identify and conduct three applied research projects on gender and social analysis to generate new knowledge on key challenges of gender, poverty and ENRM.
2 Strengthen capacities of researchers to implement action-oriented research on SAGA in ENRM through technical support to broaden current research approaches, methodologies and analysis through improved peer learning and networking among researchers in the region, and through peer exchange, joint analysis and capacity-building activities.
3 Identify key issues affecting organizational support for integrating SAGA in ENRM research through structured discussions on challenges, successes and strategies with researchers and other stakeholders.
4 Disseminate key results, lessons and methods of the program to different user groups through communication strategies and production of media and materials.

This report will describe and evaluate the results of activities undertaken under the capacity-building component (component 2), and because capacity building was seen as a cross-cutting theme, the report will also comment on aspects of capacity building related to the other project components.

The four case studies: a characterization of the institutions involved[3]

The four teams had different starting positions in terms of their sensitization to skills and practical experience with gender research. In fact, the selection of case studies included those submissions with a low prior knowledge of SAGA, because the benefit was expected to be highest. The research teams differed with regard to their prior knowledge of SAGA tools, the presence of difference skills and scientific backgrounds, and their institutional linkages (Table 13.1). The four teams also had different capacity-building needs and benefited differently from the capacity-building interventions as well as from the group interactions at the joint workshops. The teams also had different strengths and experiences to share, such as strong field research skills in Algeria, linkages to civil society and the private sector in Lebanon, and strong linkages to academia in Morocco and in Algeria. The teams brought these strengths to the project for the benefit of the other teams.

Table 13.1 Inventory of social analysis methods proposed by participants

Country teams	*Methods proposed*
Algeria INRA	Timeline Field force analysis Problem tree Resource map Socio-cultural map Individual surveys with checklist of questions Market survey
Lebanon CRTD-A	Focus-group discussions Timeline Field force analysis Problem tree Resource map Socio-cultural map Thematic maps
Morocco ENA – Meknes	Daily and seasonal calendars Historical profile Stakeholder analysis Participatory observation Resource map Rainbow diagram Problem tree The wheel
Morocco ENFI – Boudinar	Stakeholder identification Village resources map Village social maps Farming systems diagram Daily activity clocks Seasonal calendars Resources picture cards

The project's capacity-development philosophy, approaches and tools

Definition of SAGA-related capacity building in the context of this project

Regional experiences in capacity building in MENA have generated substantial lessons and insights to inform future strategies. Short-term training programs and courses are useful to improve exposure to SAGA concepts and methods. However, experience showed that such training has limited impact if it is not complemented by practical research application and field-based technical support. Researchers engaging in any capacity-building program need time for 'learning by doing' through iterative learning, field-based practice and ongoing mentoring. Hence, capacity-building efforts must be well grounded in research activities.

Capacity strengthening in this project was not synonymous with training. Rather, training was one activity towards strengthening capacity, along with other methods of 'learning by doing', mentoring, and peer learning. Capacity building was an iterative continuous process, throughout the project cycle. It began during the pre-project phase, with an extensive collaborative project preparation phase which helped define the status quo of SAGA in the Middle East and North Africa, and built a strong regional community of practice. This pre-project phase included three main activities: 1) a review of regional research and capacity-building projects (Fajber 2008, see the box below); 2) a workshop with key regional experts (Feldstein and Abdelali-Martini 2008) to validate and expand these findings and to discuss ways forward; and finally 3) a proposal planning workshop which built on these lessons and developed the framework for the project proposal and formulated the call for case study submissions.

The design of the project's SAGA capacity-building components also built on the identification of key lessons from other regions (Fajber 2008):

- limitations of short-term training and one or two lectures;
- grounding capacity strengthening in research: 'learning by doing';
- the need for ongoing mentoring;
- opportunities of peer review and exchange;
- value of local cases, and resources in local languages;
- institutional support to implement SAGA;
- investment in committed scientists (and organizations) over the long-term;
- employ multiple modalities.

The project's capacity-building activities, tools and approaches

The design of the project's capacity-building component was built on the above listed lessons derived from SAGA research undertaken in other regions. The project used a customized approach to capacity building: an assessment of capacity strengths and gaps informed the selection of the appropriate research tools, and the development of training materials. This process was greatly facilitated by the assignment of a designated field research support coordinator to each of the four regional teams. These were able to respond in a timely manner to the particular needs of the teams that they were assigned to mentor. It was understood and accepted that capacity needs went beyond SAGA skills, but included other aspects of the research cycle. Even improving the research proposals was seized as a capacity-building opportunity – research question, research team composition and roles, as well as budget needs and timeline were reviewed and revised where needed.

The main methods and activities applied in the capacity development of the project are given in the box below.

Capacity enhancement methods and activities used in the SAGA project:

1. learning workshops (international, multi-team);
2. initial training in participatory tools (SAS);
3. collaborative backstopping from international researchers;
4. field practice in using tools;
5. involvement of students;
6. team building (discussions, SAS workshop);
7. sharing with other students;
8. virtual communication (SKYPE, email, Picasa, ICARDA website);
9. graphics to conceptualize research and analysis;
10. language training (tutoring to meet special needs);
11. group discussions (for example for identification of major findings);
12. collaborative writing (outlines, review and comments);
13. peer review and comments on each other's work to stimulate discussions and peer learning (for example the Beirut workshop);
14. review and critique selected journal articles for encouraging thinking about scientific writing;
15. writing guidelines (for example those provided in Beirut);
16. poster developments and presentations;
17. oral presentations.

The main capacity development methods and activities are briefly described here.

Establishment of a technical group

The establishment of a technical group, which was part of the original plan, was slightly adjusted. The establishment of an advisory technical group had been initially envisioned to guide the research teams. This approach was soon abandoned because of timing and scheduling difficulties, and because it was found that an advisory team was not responsive enough in a timely manner and lacked the detailed knowledge of the teams' field research needs. Instead it was decided to expand the role of the field advisors in their mentoring role.

Capacity strengthening needs assessment

During an initial capacity assessment workshop the individual capacity needs of the four teams were assessed using questionnaires and group sessions. The capacity needs and priorities expressed by the teams went well beyond SAGA skills.

Development of work plan for capacity-strengthening activities

Based on the capacity-needs assessment the priorities were established and an individual workplan and timetable for capacity-strengthening activities and specific consulting needs was agreed upon for each of the four teams. These activities were then carried out during country-specific workshops, as well as regional workshops.

Development of targeted training materials for research teams

The selection of tools and training materials: it was decided to select and adopt the SAS^2 tools for the purpose of SAGA research. These tools were well-documented and tested; training materials and expertise were available which made it a practical choice given the time constraints. The potential SAS^2 tools were initially presented to the four teams at a workshop, and from the host of SAS^2 tools, the teams, in discussion with the project management, selected those tools that were considered most useful for their planned research work.

Support and mentoring to researchers in the field

Backstopping and skills development by experts: the project's capacity-development activities focused on capacity needs during the research cycle: proposal writing, research methodology definition, formulation of research questions and hypothesis, data collection and data management, design of the questionnaire (for Algeria only), data entry, data analysis, report writing and the communication of results. Each research team was assigned one field advisor. The four advisors (two ICARDA scientists and two consultants) also played an important role for regional knowledge sharing and communication across the four teams.

Peer learning and exchange

The project workshops – some of which took place during the preparation of the project, prior to the selection of project proposals – provided the main vehicle for peer learning among the four research teams and their field mentors. Two workshops were conducted during the planning and development of the project where many regional experts were involved; four learning workshops were organized during the life of the project. These workshops provided good platforms for peer learning and many of the capacity-development activities were implemented in these workshops. Finally two outscaling workshops were organized at the end of the project (Algeria and Morocco) where stakeholders from wider research and development groups were invited, results were presented and implications were discussed. The list of workshops is given in Annex 1.

The Learning-by-doing process

The following capacity development steps constitute the hands-on learning approach implemented in this project. This hands-on learning was central to the capacity-building goal of the project. The key elements of the learning by doing are described below.

Research project proposal development

The process of soliciting proposals and the subsequent process of iterative review and feedback on the proposal, the research plans and methods and further revisions of those plans were a critical approach which helped the teams to practice proposal development and contributed to the team building.

Formal training in methods

The research methods training workshops in Settat (Morocco) and in Aleppo (Syria) were designed to help the teams to formally learn specific research methods. This learning was further invigorated by ensuring that the teams selected specific methods for each research question that they proposed in their research plan.

Experience in methods application

The fact that the quality of research would be evaluated by how well the methods are applied created an incentive for the teams to understand these methods. The teams documented how they applied the selected methods and identified how they have adjusted each method into the conditions of their case studies.

Peer review of research proposals

The teams have presented their research proposals in two consecutive workshops and others have commented through virtual reviews on the deficiencies of each proposal. The feedback provided critical peer review opportunities and each team made substantial revisions of their proposals and strived to make the research questions more clear. This peer review process was used throughout the project and was also applied to the analysis and presentation of results.

Involvement of students

The teams were strongly advised to involve students in the field work as this would provide more long-term impact in capacity development in gender analysis. The two teams from Morocco included two female MSc students. Of the Algerian team one member did her PhD on the project, while the Lebanese team involved student investigators from a private university.

The Social Analysis Systems (SAS[2]) tools – their selection and assessment of success

Social Analysis Systems are concepts and tools for collaborative inquiry and social engagement that lend themselves very well to be applied to gender assessment and gender research. These participatory and action research tools are the cornerstone of the capacity-building efforts of the project – they are the key set of additional skills that the project was intended to promote. The basic research skills the project had built prepared the teams for the successful application of these social engagement skills in the context of their research proposals. The methodologies are described fully with field applications in the area of NRM and agriculture (the toolkits are online in Chevalier and Buckles 2008[4]). Introducing and choosing these tools and applying them to the gender research questions was a reasonable and also a practical choice in the context of this project. The SAS manual brings together some 80 tools for the purpose of participatory social research, a selection of which were presented to the teams at a project workshop. Then it was the task of the research teams to choose a set of tools that would enhance the quality of their research and facilitate learning for change. The tools applied in this study are a selection based on extensive group discussions that resulted in four SAS tools being chosen for mandatory use by all four research projects. These were causal dynamics, time line, force field, and social domain.

The usefulness of the SAS tools was measured on the basis of their ability to illicit accurate information that could be used not only by researchers, but that would also contribute to enhancing social capital within the community.

Criteria used for the assessment of the SAS tools

The objective was to contribute to an understanding of the usefulness of SAS tools from a gender perspective by comparing the comparative appreciation of women and men through 1) ease of use, 2) satisfaction with the information generated and 3) their capacity to enhance social capital.

Ease of use and control of the tools

To assess the level of control over the tools (comprehension and manipulation), the teams observed the men and women in their research sample using the tools. To measure comprehension, they used the number of reiterations the subjects required to understand the tool before using it. They observed men and women during the sessions, identifying those who had difficulty (for example they asked others for explanations). It was found that the comprehension and use of each of the tools differed by gender.

Satisfaction with the information generated

The tools were also evaluated according to the amount of satisfaction (high, medium, poor) participants felt with the information generated as a result of using them. Satisfaction was measured on the basis of two criteria:

- the perceived usefulness of the information for the future (for example to solve problems, to improve a situation);
- whether the experience of generating information led to something new (for example a new perspective on their daily life, organizations in the community);

Contribution to the enhancement of social capital

The effects of the four tools on the construction of individual and collective social capital were measured using several criteria. The criteria for individual social capital included 1) if self-esteem was increased, 2) if each individual was recognized by the group, and 3) if trust was built between members and non-members and the organizations, between members of the same organization and of different organizations, and between non-members (other people in the community). The measurements for gain of collective social capital included a) trust established between organizations and with the wider community, b) cohesion of the group or organization was improved, or c) new networks were created.

As a next step, it would be beneficial for future gender research if the SAS^2 tools could be 'engendered' and developed into SAGA tools with the research question examples from this project, and be made available online, labeled clearly as SAGA tools.

The evolution of the research proposals and research methodology

An iterative, hands-on peer learning experience in capacity building for proposal writing

The iterative capacity-building approach started with the improvement of the originally submitted research proposals. The peer learning experience in proposal writing led to some revisions and a successful evolution of the submissions: the research questions were further shaped and focused to bring out the SAGA aspects, which also enabled identification of the capacity needs of the teams. Through this hands-on process, the teams also acquired important proposal writing skills that will benefit their subsequent research work and collaboration.

For example, in Algeria the initial proposal evolved from a broad, overly-ambitious natural science-based approach to a well-focused work plan with clear methodologies including gender research tools. Rangeland and livestock scientists at INRA in the southern Algerian Djelfa Governorate, had developed the initial proposal. Its goal, '...maintaining the different potentials of the resource to ensure both the preservation and sustainable use...for the maintenance of production systems and to improve the living standard of poor households...particularly for women', was very general. Furthermore, it had eight different objectives ranging from growing knowledge about environmental resource degradation, vulnerability, climate change; crops and livestock technologies, breeding; strategies to mitigate vulnerability; development of a glossary of local terminology; a database of traditional techniques related to livestock production; the establishment of a conservatory of traditional heritage of the Djelfa region and the analysis of policies and their impact on perennial resources in the region. Furthermore, the methodology proposed a multidisciplinary approach for understanding the drivers of change, a comparative approach between different systems, and the use of community participatory and qualitative approaches, without having a clear understanding of the specifics of these approaches or their application. The proposal also suggested a link with a development project in the area without identifying clear linking options.

Through the capacity-building activities the research questions were refined, and the capacity of the team in social and gender analysis was built through hands-on research examples on the use of engendered participatory and social investigation tools. The proposal evolved into a well-focused research proposal with four distinct and complementary outputs (deterioration of natural resources and their impacts on women's work; the loss of know-how and its impact on women and girls' work; the access to market for a specific garment; and occupational health issues related to the use of wool and camel hair) with a clear methodology and a specific tool to answer the research questions.

In Morocco Boudinar the initial proposal was written without a social scientist on the team, and lacked a clear focus on women, as well as other stakeholder involvement. The proposal was broad and ambitious it had planned the organization of women into associations, and the design of income-generating activities for women, and the training of women in water and NRM practices. The planning workshop and subsequent iterations of adjusting the proposal brought out the gender dimension more clearly and helped to focus on achievable objectives and activities.

The main adjustment in the Morocco Maamar proposal was in the reduction and prioritization of objectives that could realistically be obtained within the lifetime of the project, and in the selection of appropriate research tools from a long list of potential participatory and social science tools the project team (which included a social scientist and gender specialist) had initially listed.

In the case of the proposal from Lebanon, the proposal was adjusted to focus less on desk reviews (of laws and regulations regarding cooperative rights). The gender mapping of institutions and of the agro-processing industries had to be scaled down significantly. Also, the expected outputs and outcomes (i.e., articles, papers, media releases), had to be revisited and adjusted to a realistic number.

Refinement of research methods

The process of focusing and fine tuning the research proposals was followed by the task of matching and selecting the most appropriate research methods to answer the research questions. Initially the team proposed between six and seven different tools for their research but this was found to be ambitious and unfocused. It was a capacity-building challenge to refine this list and chose the appropriate methods for the specific research questions, and to then build the capacity for the appropriate use of those methods. The teams reduced their choice of methods to a maximum of four methods that were relevant to their specific research question.

The research teams' own assessment of the project's professional and organizational impact regarding SAGA capacity

The starting position of the four teams in terms of their institutional strengths and the individuals' research skills varied from team to team. The project's approach was to tailor capacity-building activities as best as possible to these individual needs. At mid-term stage feedback was sought and used to evaluate the progress of the capacity-building activities, and to adjust actions where this was necessary. At the end of the project the teams were asked to evaluate the effect the capacity building had had on them professionally, as well as on their institutions.

Project participants' capacity self-assessment at mid-term stage and adjustment action

At mid-term review stage the project took the opportunity to receive feedback from the four research teams on what had been achieved so far, and where there were still gaps, or a need for adjustment. This was done formally through a survey, as well as in working group discussions. The results were used for course corrections and refocusing of support activities. This flexibility to adjust the work plan and the tailored approach is a clear strength of the project approach and benefited the four research teams.

During the mid-term workshop, working groups discussed the following question: what capacities do you have now that you did not have at the beginning of the workshop? What have you learned? The list below summarizes the main responses:

- the importance of adapting tools and research questions to the cultural context;
- gained appreciation for the richness of information provided by different social groups in field research;
- learned with the local community new ways to use and adapt tools;
- increased understanding of gender analysis;
- used SAS tools in the field to answer gender questions;
- identified new research questions;
- learned to use and compare different tools methodologies for gender analysis;
- communication skills are needed for qualitative research;
- how to design projects and formulate a research hypothesis with more clarity and precision;
- to write-up research projects in a clear, concise manner, for easy implementation;
- the importance of English as a working language;
- improved research team management;
- better understanding of the importance of inter- and multidisciplinarity;
- built a strong network among the four research teams.

The above list shows that already at the mid-term stage of the project a variety of soft skills, field knowledge, an appreciation for gender work and its results, as well improved research skills had been built.

At the mid-term review meeting the teams also analyzed the capacity-building modalities – how the project enhanced the capacity of the research teams through workshops, field support, information sharing and collaboration.

The workshops

The workshops were seen as very useful to improve the understanding of SAGA concepts, to continuously revisit the research questions, in receiving peer feedback in the process of having to defend the proposals in front of the other teams, and in selecting the methodologies that are best suited to provide answers to the research questions.

Field support teams

The majority of respondents found that the field support teams helped in understanding concepts and implications of gender in NRM, in the adoption of SAGA methods during fieldwork and in the analysis of the implications of gender in NRM.

Information sharing and communication

Information sharing and communication during the workshop events was seen as satisfactory, but the teams felt that improvement was needed in communication and collaboration among project partners within the teams as well as among the four country teams.

The mid-term evaluation also invited ideas for improvement and further adjustment of the capacity-building activities. The main suggestions were:

1 Improve communication among project teams and with ICARDA.
2 Apply the SAGA tools with support in the field.
3 Explore new tools and adapt to findings.
4 Develop a manual for engendered SAS tools adapted to NRM research questions.
5 Involve people from outside the project (local communities, other research institutes).
6 Receive continuous support with scientific writing and proposal writing skills.
7 Compare research methodology and experiences among the four countries.

The project management and the field advisors reacted to these suggestions by focusing the capacity building on these needs and adjusting time tables and rebalancing activities accordingly.

Evaluation of the effectiveness and impact of the capacity-development effort on the participating teams

The effectiveness and impact of the project's capacity-building approach on the participating country teams has been evaluated based on a self-assessment at the end of the project, as well as based on reflections by the project management and implementation team.

Project participants' self-assessment at the end of the project

Through their participation in the SAGA project, the project partners participated in a number of discussions and specific training sessions related to gender-sensitive research capacity. At the end of the project, the partners were asked to evaluate the immediate and lasting effect their participation in the project and the specific gender training courses had on them and their professional environment. This was done in person during the final workshop events, by phone or by email.

1 What effect did the gender capacity building have on your own professional career and behavior?
2 What effect did the gender capacity building have on your organization (for example, with regard to research topics, partnerships with other organizations, fund raising, curriculum change, degree planning, other)?
3 What lessons have you and your organization learned from participating in the SAGA project regarding gender-capacity building, and how would you apply these lessons in a similar future project context? What would you do differently, and what has worked well?

Below are some examples of the country-specific experiences and evaluations.

Algeria

1 The combination of the new interview tools and the existing expert knowledge of the INRA researcher of pastoral systems is very beneficial. The team learned that gender alone is not the topic of the research, it needs a research question entry, to which then the SAGA tools are applied.
2 For the Algerian team the capacity building has improved their proposal writing skills which were then put to use. The research results from this project informed these new proposals; a gender dimension is properly integrated. New partnerships with the National Institute of Agriculture and the department of socio-economics were formed through the interdisciplinary approach. Through these new partnerships a research module on gender research at the institute and a PhD student from the Algerian team with a topic on gender and NRM is now co-supervised by the socio-economics department and ICARDA. Through the team's research, for instance the occupational health issue, the team and the professor have realized how important the gender

dimension is in identifying issues that have the potential for changing national policies (see Algerian case study on occupational health issues, Chapter 5).

3 Government programs in support of women artisans and the survival of their craft have been devised and funding for micro projects in support of women has been planned. The project induced debates about the role of women in society and the household. Research and academic linkages have been formed and strengthened, an introduction of the gender research curricula in the students' programs is planned.

Lebanon

1 It was inspiring to see the empowering effect the participation in the research had on rural women.
2 The NGO plans to work on refining the SAS^2 tools to become gender analysis tools.
3 Partnering with a private university for the participation of students in the project through a co-academic program was successful, and it was a good exposure for the students, it improved their field research and interview skills. The NGO will use the results of the work in its advocacy work, and has engaged with the private processing sector to raise awareness among factory women for the discrepancies between women factory workers and coop members. They also shared the results of this study at the CSW (Commission for the Status of Women) forum in New York last March.

Morocco Boudinar

1 Collaboration with an international organization and with regional teams was a new and enriching experience.
2 Established an email network with the institutions and stakeholders involved and created an online discussion group. Women's peer learning was practiced and found to be very effective in learning and awareness building. The experience of working in the project made the team more open toward including other stakeholders like NGOs, cooperatives and local community organizations in project work, and their contributions were found to be enriching.
3 The team found that the exposure to new analytical tools has even helped them to analyze past development efforts and understand failed efforts to benefit women. They have learned through the training events why previous programs were not sufficiently inclusive in addressing gender issues, and because of their knowledge of tools and approaches corrective action can now be taken.

Morocco Maamar

The implementing agency is mainly a national educational institution where most research is done through student projects. The capacity-building activities included the introduction to and use of participatory tools, field research guidance, the formulation of research questions, report and scientific writing skills. The team benefited from the exposure to English as a working language.

The impact of the 'means of delivery' of the capacity-building efforts

The quality of the research proposal submissions, as well as the assessment of capacity needs that was carried out at the beginning of the project revealed that the capacity needs of the four teams went beyond SAGA skills, and included research methodology, data management and analytical skills, as well as soft skills related to writing and communication. To focus the capacity-building efforts only on building SAGA skills would have not been sufficient. The broad efforts to improve the research process by addressing these other capacity needs through the close customized guidance of field advisors improved the overall research quality. It was understood from the beginning that improving a specific set of skills, such as SAGA skills, cannot be done in isolation, but that it needed to be embedded in a sound research planning and analytical process.

The three main vehicles of delivering capacity building were:

1 Interactive processes of capacity building throughout the research cycle (improving the research process by building capacity in formulating research questions, field research methodology, interview guides, data analysis, reporting skills, scientific writing).
2 Workshops that included the entire project teams and key stakeholders; at which there was regional knowledge sharing and building of a strong community of practice.
3 Close guidance was given through field support teams, including offering student mentoring.

Time and resources were spent by the project on an interactive and responsive process to improve overall research quality and skills, even if perhaps this took away time and resources that could have otherwise been applied towards more details of applying SAGA tools. Responding to the broader capacity needs strengthened the research teams' capacity and will enable them to apply the SAGA skills more effectively in future research and in formulating future project proposals.

The project workshops

The project workshops were inclusive and hands-on: they included the entire project teams, as well as key stakeholders, and were designed with a lot of time for feedback sessions and interaction, with less emphasis on presentations. Even when presentations of preliminary research results were given, the group provided constructive critique and feedback, turning the presentation into a learning opportunity for all to enhance scientific writing skills and seek clarity of arguments.

Field support advisors

The guidance given to the four research teams throughout the project cycle by designated field support advisors was perhaps a luxury but had a high return in terms of quality enhancement, because of the flexible and timely manner in which the advisors were able to react and assist. The prior exposure of the field advisors to gender research was an obvious benefit, but their effectiveness in this project context was also due to their broader research experience and research planning and soft skills. Even where teams at first were not used to having such hands-on advisors, the value of their interventions was quickly recognized and appreciated.

The involvement of graduate students

The involvement of a total of eight female graduate students in the four research projects is an indicator for a long-term effect of the capacity building, as these students started their career with exposure to SAGA skills, and to field research more generally. The involvement of graduate students in delivering part of the research agenda also enhanced the linkages between research and academia. In Morocco and Algeria Masters and PhD students were used by the project to carry out field work for their graduate projects and in Lebanon, graduate students helped with interviews and data analysis and write up. In all cases the students and their teams benefited from the SAGA tools training, as well as from the opportunity to then apply the theory to actual field work. The model of using students and embedding field interviews in graduate projects might be repeated by the institutions involved, tightening the linkages between research and academia, and perhaps even allowing for a feedback loop from the research institutions informing the universities how the current curriculum could be adjusted to prepare students better for gender-related field work.

Tailored tutoring from a multidisciplinary team

The quality of the research work benefited from the hands-on tailored tutoring on all aspects of the research cycle, and from the positive experience of working in a multidisciplinary team where each profession contributed to the outcome. This will have lasting effects as it may lead to the production of better research proposals, with a better chance of receiving research funding, especially when the research proposals are formulated with a gender focus and methodology.

Peer learning

Participation in the project resulted in the acquisition of new skills. This achievement was not just from the exposure to the SAS/SAGA tools, but the general research and collaboration skills like peer learning in a regional project that the project developed. Capacity development contributed to scientific writing and proposal writing skills, data analysis and presentation, and, important especially in the North African context, English skills.

SAGA skills were seen as complementary to existing skills. The skill sets inherent in the teams at the beginning of the project were seen as important assets upon which the additional SAGA skills were built. Strong research skills, field experience, and the knowledge of local institutions and conditions were appreciated and successfully complemented by the new set of collaborative and social inquiry tools. The mix of skills and backgrounds inside the teams and the collaborative nature of the project activities fostered multidisciplinary working.

The generation of new knowledge

This is reflected in the number of papers and journal articles that were written or are planned to be published on the new insights and answers to the formulated research questions. These are in part dealing with constraints faced by women in NRM-related topics, their role in providing labor, and in holding the knowledge of traditional crafts making. The knowledge generation is also an indication of the applied nature of the capacity-building efforts: it was not training for training's sake, but SAGA capacity was built to enable the teams to carry out better quality and more relevant research activities.

The more general capacity efforts towards better research skills improved the overall quality of the outputs.

A total of eight theses, four research papers in Algeria, five in Morocco, three in Lebanon have been produced. A discussion of the role of the SAGA methodology in obtaining the research results in comparison with past research methodology is an important part of every one of these papers.

Raised professional profile and exposure

Participating in the project with international research and development partners raised the professional profile of research team members and that of their institutions. The capacity-building activities exposed the team members to innovative novel ways of carrying out participatory research, interact with other regional and international research and development organizations in workshops and through collaboration on journal articles and future research proposals.

Lessons learned for future activities

Intensive support for a small number of teams

Supporting a limited number of proposals through personal, intensive, hands-on tailored capacity building throughout the entire cycle of a research project proved to have lasting impacts on those research teams – new institutional cooperation and the application of gender aspects and methodologies in new research proposals are indicators of this success. With the same level of funding the selection of a larger number of proposals would have resulted in a reduced ability to tailor the capacity-building activities, and less attention to detail at workshops, and less time for communication and interaction among the teams and with the resource teams.

Selection process of initial research proposals

The research proposals were chosen based on their potential to show the added value of SAGA research tools and capacity, not based on high levels of prior knowledge of gender research or the inclusion of social scientists on the teams. The selection was based on the interest of the research teams to work on improving the gender focus of their research. The appreciation expressed for the added value of the SAGA tools in the evaluation at project end show that this philosophy of supporting proposals with weaker prior knowledge has been successful. The project looked to support champions in their willingness to learn, not for experts keen to apply their knowledge.

Success through self-sufficiency

The project did not rely on outside SAGA experts to carry out the research work, team members had to work through the project cycle and apply the capacity-building lessons. And while the research support teams assisted and worked with them closely, the actual field work and data analysis was done by the research teams themselves, often with the help of graduate students.

The added value of interdisciplinary team work

Working in interdisciplinary teams was a new experience for many of the teams. All disciplines added value – the project was able to demonstrate the added value of social scientists with their prior experience of participatory approaches, and the natural scientists had an in-depth knowledge of the study areas and the natural resource situation, as well as field work experience. The new SAGA research tools research skills were embraced by both sides, and seen as adding value to solving jointly-formulated research questions.

Building on regional capacity and team work

The sequence of regional workshops provided ample time for interaction, feedback and discussion of institutional and professional experiences which was conducive to reflection, regional learning and the building of professional linkages. Some teams already had closer institutional relationships between academia and research, some were already accustomed to the use of graduate students, for some the cooperation of social and natural science researchers was common practice – working in these various contexts was an enriching learning experience.

Language barriers remain a challenge

The use of English for the dissemination of the research results and the communication of the experience of working with SAGA approaches remains a challenge for the Francophone countries, especially for the Moroccan and Algerian teams.

Develop engendered SAS² tools

The adaptation of the SAS² tools to the gender research context proved to be natural and useful. Documenting the adaption of SAS² tools for the gender research context using the examples of this project, and then making them publicly available online through toolkits would be an ideal follow-up action to this project. It would be a welcome next step that all four teams expressed a desire for at the end of the project.

A dedicated role for information sharing

Internal and external information sharing and communication are crucial for the success of any project, but this is especially important in a regional project, where several country teams are involved, and for a project such as this one, where the focus is on capacity building. Internal communication is needed to keep all country teams involved and informed, and to facilitate the sharing of experiences with each other (including, in the case of this project, translation of workshop proceedings and papers into English). The external role of communication is needed to reach diverse stakeholders. The research teams often lack these specific communication skills, and they lack the time to carry out this function, and cannot be expected to play this role. Planning and budgeting for the inclusion of a dedicated and skilled staff member would benefit the outcome of the project.

Potential for curriculum change

The close institutional linkages and collaboration of research institutions with academia that were forged through the project or had already been present especially in Morocco and in Algeria bear the potential for curriculum change and the inclusion of additional courses and content. The project showed that skills related to field research and gender analysis are in demand, as well as soft skills such as proposal writing, scientific writing, communication skills, and English language skills. The close institutional linkages between research center and university also present the opportunity to formalize the use of graduate students in field research projects, for an early exposure to and practice of academic training related to gender research.

One could even imagine a more proactive approach from the donor/IDRC side to actively engage with academic institutions as a target group and sponsor Masters' Degree level curriculum enhancement, in the area of SAGA, but also in much needed soft skills to train the future generation of research leaders.

Notes

1 Consultant at the World Bank (temporarily located in Amman).
2 International Center for Agricultural Research in the Dry Areas (ICARDA), Amman.
3 The case studies are detailed in Chapter 1.
4 www.idrc.ca/EN/Resources/Publications/Pages/IDRCBookDetails.aspx?PublicationID=108

References

The chapter draws on project progress reports and workshop summaries and presentations prepared during the lifetime of this project.

Fajber, E. (2008) 'Integrating Gender and Social Analysis in Environment and Natural Resource Management Research in the Middle East and North Africa: A Review of Challenges and Strategies'. Unpublished report for IDRC and ICARDA.

Feldstein, H. and Abdelali-Martini, M. (2008) 'Strengthening Regional Approaches to SAGA in ENRM in MENA: Expert Consultation Workshop' (9–10 April 2008); summary report.

Annex 1: The learning workshops organized in the SAGA project

1. Strengthening Regional Approaches to SAGA in ENRM in MENA: Expert Consultation Workshop. Aleppo, Syria, 9–10 April 2008, where SAGA experiences from other regions was shared.
2. Strengthening SAGA in ENRM in MENA: Toward More Effective Research and Development Proposal Planning Workshop. Cairo, Egypt, 19–21October 2008.
3. Training Workshop on SAS[2] Concepts and Tools to Collaborative Inquiry and Social Engagement organized by INRA. Settat, Morocco, 26 November 2009, 18 participants.
4. Proposal Development Workshop. Settat, Morocco, 7–8 November, 2009, 18 participants.
5. Gender Analysis Methods and Research Implementation Workshop, ICARDA. Aleppo, Syria, 2–6 May 2010, 19 participants.
6. Training and Follow-up Workshop. Beirut, Lebanon, 28–30 September 2010, 20 participants.
7. Training and Follow-up Workshop. Rabat, Morocco, 16–19 May 2011, 20 participants.
8. Two Workshops to Scale-out Research Results and Methodologies, to share results and plan future work. Algiers, Algeria and Rabat, Morocco, December 2013, about 70 participants.

14 Conclusion

Aden Aw-Hassan[1] *and Malika Abdelali-Martini*[1]

This book presents a research and capacity development experience in MENA region. The project, in which the presented research was implemented, was formulated on the hypothesis that among the reasons for the very slow development of gender research capacity in the MENA region is paucity of local reference of research experience. As a result this project had two main objectives: 1) to conduct case studies of gender research in order to generate local experiences for reference in the region and collate them in the form of a reference book and 2) to build the capacity of research teams in gender research. The results of these two main objectives have been presented in this volume.

The project applied two main methodologies: 1) In the research, participatory research methods and tools were selected from the SAS[2] Guide to Collaborative Inquiry and Social Engagement (Chevalier and Buckles 2008) and applied in the field. The analysis has mostly been qualitative. 2) For the capacity development during the lifetime of the project, we deployed an iterative process of training, field applications of methods, further training, mentoring and support by a technical support group; and validation of the data through bringing it back to the research participants.

The research in Djelfa, Algeria, confirmed that there is a strong relationship between the degradation of productive natural resources (rangelands) which is affecting grazing patterns of livestock, and the consequent drop in the supply of raw material, and the loss of local knowledge related to women's production of a traditional garment, *qashabiya* (Chapter 2). The results highlighted how the environmental changes affected women's decision-making regarding the organization of labor within families in terms of the use of different types of fibers to weave *qashabiya*. In addition, there is clearly a lack of interest among the youth to take over this activity from the elders. Results also show that women produce these valued local garments but only partially capture their value as a result of social constraints on women's participation in the market place. The findings show that despite important economic changes and modernizing the textile industry at the national level, the *qashabiya* market remains heavily compounded by the traditions excluding women from selling their products at the market place, thus having the opportunity to bargain over the price of their products and earn a good income from their labor. But rather those restrictions give (male) intermediaries and traders a free ground to control *qashabiya* marketing, leaving little control and bargaining power for women producers. This disparity in market access makes male intermediaries the main beneficiaries (see Chapter 4). The low earnings from this craft affects women's interest in this traditional work, particularly the girls, and count among the major constraints on the sustainability of this national heritage. This research also investigated women's occupational health resulting from *qashabiya* production (see Chapter 5). Results indicate that women are affected by a wide range of ailments associated with their age, raw material processing activities, weaving and other socio-economic factors, and that this sector of activity is potentially placing women and children at risk of serious illness, as they are not covered by the national insurance plan due to the informal and home-based nature of their activities. Furthermore, the sector has not developed for many decades most probably because the activity is mainly performed at home, thus hidden from development initiatives. Many women and also medical doctors in the area are unaware of the relationship between women's illness and their activities. Improvements to the working conditions are discussed, and the research identifies also some areas where technology interventions can ease women's working conditions. It is important to elevate this activity to the national level: it should be considered a formal activity covered by national laws that would contribute to changing the gender relationships and power of this activity.

The research on the market access of rural women's cooperatives (RWCs) in Lebanon (Abou Habib, Chapters 6 and 7) confirmed the observations drawn from daily practice, namely that material capital investment can only be useful when there is day-to-day engagement with rural women to support the development and building of a battery of technical and inter-personal skills, understanding and improvement of intra-household gender relations and the ability to understand and subsequently address obstacles that block market access. There are some top-down large and costly aid programmes which have presumably diverted the attention and capacities of rural women from looking at and addressing marketing issues throughout their diversified efforts. The tools used in this research allowed the construction of a timeline of research on women's cooperatives and draw, as a result, an interesting pattern of international funding and its impact on women and their small enterprises.

Analysis of the male-dominated food industry through case studies generated lessons for the rural women's cooperatives (see Abou Habib, Chapter 7). The analysis shows that activities of RWCs are generally small in size and have low production capacities, have low financial means, but have women with high production skills and strong motivations as leaders and decision-makers in their enterprises. The study provided strategies for improving market access and gaining steady market share, among these are: sustaining quality and consistency, linking with agent distributors and attractive packaging. The study also noted that rural women's cooperatives have a strong comparative advantage over the private sector, namely in the quality of their production and the nature of their enterprises.

The case study in Boudinar, Morocco (Khattabi *et al.* Chapters 8 and 9), provides a detailed description of gender division of roles and tasks in a remote, conservative, rural society. The effects of and resulting vulnerability from climate-related events (droughts, water shortage, floods and landslides) by gender and particularly on women's workload are analyzed. The analysis provides insights on how environmental factors can further exacerbate gender inequality, making rural women particularly more vulnerable, and thus calls for redoubled efforts in regions that are more vulnerable to climate variability. This research shows how men and women farmers cope with climate variability, and explores how coping strategies are shaped by gender power relations and the decision-making process. At the same site, a 'case study' approach was applied involving 28 individual women of different socio-economic levels and documenting their day-to-day situation under environmental and socio-economic challenges. These case studies support the call that for any agricultural effort aimed at benefiting both men and women in the rural commune of Boudinar, gender differences need to be understood in their complex, real and dynamic life situations.

The second Moroccan case, in Maâmar, (see Chapter 11) presents the case of a large development project which had not sufficiently integrated gender in its design and how that could have limited the success of the project. The development project in the community was to implement sustainable utilization of herbal and medicinal plants for community income generation. However, although the project supported women's groups, the project, for reasons not fully explained, did not fully integrate gender in the implementation of its major development initiative. The dichotomy of supporting women's groups but not fully integrating them in the major development initiative was surprising. This study documents the importance and role of women in the herbal and medicinal plant industry, and demonstrates that women play a major role in the utilization of and in maintaining knowledge for herbal and medicinal plants. The results illustrate that lack of sufficient gender integration in local community development initiatives is a program design flaw which could result in a failure of such initiatives.

In the same study, researchers (see Chapter 10) show that during the application of participatory research methods men and women have different capacities to participate due to their level of education, exposure to external ideas and ability to express thoughts. Women had greater difficulty in applying some tools. These analyses show that when designing implementation of participatory research tools, the factors that may affect the level of participation of both men and women should be considered and remedies should be made to avoid incomplete participation. This study also examined the extent to which local participants found the results generated by the participatory process useful. This raises the point that one should not use participatory tools only for extractive purposes but rather should use them for transformative purposes by providing space for participants to reflect and think about making changes on their own, while considering possible socio-economic factors and gender differences in the level of participation. This would be the first step for transformative gender research in the form of participatory evaluation of interventions after the diagnosis phase is completed.

One of the studies, at the Boudinar site, Morocco, (Chapter 12) investigated the effects of the participatory research on the local community in terms of awareness of and aspirations for better interactions with external agents. The research results recognize that participatory research is the start of a change process. Furthermore, the participatory activity itself as factor of change deserves a solid conceptualization, planning and action, and then the effects of the process, in terms of changes in participant attitudes and actions, are equally as important and should be observed. They are the essence of the research process and therefore deserve due attention and are certainly worthy of recognition and wide dissemination.

Recommendations

1 Develop equipment and improve women's market access

For rural women making *qashabiya* garment in Algeria, the facilities that meet artisan industry standards and women's needs should be developed; women's market access should be improved by empowering them through strengthening the capacity of their organizations, providing a special section exclusively for women sellers, institutionalizing women's associations, and providing market information. These two aspects will encourage the younger generation to sustain this local knowledge and at the same time earn a reasonable income from it. The value of this local income-generating traditional activity can also be improved by giving the clothing the status of protected indigenous product. The occupational health of women making the traditional garments (*qashabiya*) in rural Algeria should be further investigated and policy and development provisions should be instituted urgently. One point worthy of note is that these artisan activities need to be recognized as proper activities eligible for the national health insurance system so that women can use the health services.

2 Governmetnal and non-governmental organizations have a responsibility to build the capacity of women's cooperatives

From the results of the Lebanon case, we recommend that governmental and non-governmental organizations and donor agencies should focus on building the capacity of women's cooperatives for governance, management, marketing and enterprise technologies that would enable them to enhance their performance and benefits. This needs an extended period of engagement, mentoring and support until they develop sufficient skills and mechanisms for operating independently. In addition, the government needs to put in place policies that are conducive to gaining market access and to the sustainability of these women-owned and managed entities. The barriers to rural women's cooperatives' access to financial resources should be removed by establishing a dedicated credit line and women's networking capabilities should be strengthened for self supporting and knowledge sharing.

3 Development organizations must support remote communities, especially the women, vulnerable to climate variations

The findings of the Boudinar case, Morocco, recommend that development organizations should reach out to the remote communities that are highly vulnerable to environmental degradation and climate variations, and because women are particularly vulnerable they need specific attention. Interventions needed include improved access to water and energy supply and to basic services of health and education. They also need information on how to strengthen their adaptive capacity to the changing climatic factors which are so visible in their areas. This should include water and soil conservation techniques, information about crop choice, livestock production and protection from floods. Another finding indicates that participatory research is the first step of change and researchers should consider a shift to adopting transformative approaches rather than extractive approaches.

4 Gender considerations are essential for any rural development program design

Results from Maâmar case, Morocco, emphasize that rural development programs should by all means ensure that gender in its local dynamics and social context is an integral part of the program design and without that the program is most likely to fail. In addition, the application of participatory research methods should consider the capacities of men and women to effectively participate. The study found that the effective level of participation is determined by gender-sensitive characteristics, such as education level and exposure to external development activities, and therefore selection of different participatory tools should consider this to ensure effective participation of men and women in research for development.

5 Gender research investment required on three levels

Capacity building recommendation: Gender research capacity development in the MENA needs sustained effort in three fronts: 1) at the academic level: investment in improving and modernizing the teaching curriculum of agriculture and development programs, 2) at individual cadre capacity level: investment in increasing the flow of graduate students conducting their theses on gender issues and moving to their professional lives with new tools and methods and 3) at the organizational level: investment in efforts to transform and update the thinking and practices of organizations to embrace gender balance, equity and equality in development.

We also recommend that development organizations should link women's organizations with young and mature researchers and graduate students to undertake in-depth gender transformative research in a participatory manner with these organizations to generate more relevant, actionable results.

Note

1 International Center for Agricultural Research in the Dry Areas (ICARDA), Amman.

Reference

Chevalier J. and Buckles, D.J. (2008) *SAS²: A Guide to Collaborative Inquiry and Social Engagement*, IDRC with Sage, Thousand Oaks, CA.

Index

For Product Safety Concerns and Information please contact our EU
representative GPSR@taylorandfrancis.com
Taylor & Francis Verlag GmbH, Kaufingerstraße 24, 80331 München, Germany

www.ingramcontent.com/pod-product-compliance
Ingram Content Group UK Ltd.
Pitfield, Milton Keynes, MK11 3LW, UK
UKHW020015200926
475702UK00014B/1899

* 9 7 8 0 3 6 7 1 7 3 3 5 7 *